T0181312

Data Analytics in Professional Soccer

Daniel Link

Data Analytics in Professional Soccer

Performance Analysis Based
on Spatiotemporal Tracking Data

With forewords by Martin Lames and Hendrik Weber

 Springer Vieweg

Daniel Link
Munich, Germany

This book is a part of a post-doctoral thesis (Habilitation) for receiving venia legendi at the Technical University Munich (TUM), Germany.

ISBN 978-3-658-21176-9   ISBN 978-3-658-21177-6   (eBook)
https://doi.org/10.1007/978-3-658-21177-6

Library of Congress Control Number: 2018934626

Springer Vieweg

Printed on acid-free paper

This Springer Vieweg imprint is published by the registered company Springer Fachmedien Wiesbaden GmbH part of Springer Nature
The registered company address is: Abraham-Lincoln-Str. 46, 65189 Wiesbaden, Germany

Foreword

By Martin Lames

Performance analysis in sports as a scientific discipline has been transformed by two recent trends. First, the advent of a new 'culture' of performance analysis in American professional sports, termed *Sports Analytics*, blurs the difference between scientific approaches and the perspectives of club management, media, and business. Being largely unnoticed up to now in Europe, and mostly restricted to the big five US sports, globalization no longer allows us to ignore these developments. Second, the potential of performance analysis methods shows no sign of slowing down its increasing impact on scientific as well as practical applications. Having just learned to adequately deal with position data extracted from video streams via pattern recognition, we must now deal with compatibility issues of this data with global and local position measurements. We will be confronted with new challenges brought on by inertial measurement units attached to sensors and by biosensors that will afford us insights far beyond measuring heart rates.

This book presents a collection of performance analysis studies conducted by the author that represents the state of the art for soccer analytics and contains a variety of actual studies representative of the type that will keep us busy in future. The meaning and impact of sports analytics on soccer data analytics is described and scrutinized. The studies themselves have in common their excellent representative samples from German professional soccer made available by close cooperation between the author and German professional soccer league (DFL). Taken together, the book and the studies included demonstrate the future of performance analysis in finding meaningful solutions to theoretical as well as practical questions based on modern data analytics.

Dr. Martin Lames is professor for performance analysis and sport informatics at Technical University Munich (TUM) and holds the Presidency of the International Association on Computer Science in Sport (IACSS).

Foreword

By Hendrik Weber

The German professional soccer league (Deutsche Fußball Liga DFL) operates one of the most comprehensive live data acquisitions in worldwide sports. Since the beginning of the 2011/12 season, a central database has been storing events and spatiotemporal data of all Bundesliga and Bundesliga 2 matches which it delivers to clubs, licensees, partners and customers. As an innovation leader in the field of sport media and data technology, the DFL is aware of the scientific and economic potential that lies in this data and cooperates with universities and other scientific institutions in finding new ways to use it.

I know Dr. Daniel Link from the beginning of my time at DFL. He has supported our organization in terms of sports science and data analytics, especially concerning definitions and consistence of observational systems and the establishment and evaluation of quality standards. What I particularly remember are the lively discussions about the definition of tackling and the calculation of passing statistics. We often had to balance different market interests, while maintaining scientific integrity and relevance for performance analysis at the same time. Daniels' contributions were an important resource and means of orientation.

This book includes the largest collection of performance studies based on data from the Bundesliga to date. Daniel presents innovative approaches to data analytics in soccer that might help to develop new products for improving match analysis in clubs and media usage of match data. Additionally, the studies provide background knowledge for youth development and education of coaches. Everybody interested in the 'beautiful game' on a scientific level will find inspiration in this book.

Dr. Hendrik Weber works as a managing director at DFL. He is responsible for game data and sports technology in German soccer Bundesliga.

Table of Contents

List of Figures

Abstract

This book explores how data analytics can be used for studying performance in soccer. The objectives of sports analytics are varied and are differentiated according to interest group. Professional sport uses the data for game analysis, training load management, injury prevention and to support player transfers. Media companies enrich their reporting with game data analyses, while data companies use sports data as a demonstration domain for the performance of their data analysis products. Sports science can use the enormous data pool to analyze performance structures, to investigate academic claims and to develop new paradigms in the development of theory. On the methodological and technological side, concepts such as data mining, machine learning and big data are becoming increasingly important.

The six individual studies in this book are based on spatiotemporal data from the German Bundesliga and were published in scientific journals in 2016 and 2017. The first two studies show how tactical structures in spatiotemporal data can be recognized and used for game analysis with the help of geometric models. The third study uses geostatistical methods to investigate the dependence of various free kick parameters on the position of the free kick. Two further papers deal with the influence of the ambient temperature and the league table position on running activity and actions of players. The sixth study evaluates the effects of the introduction of the free kick vanishing spray. The findings might help coaches to improve the performance of their players and inspire other researchers to advance sports analytics.

1 Introduction

The digital transformation is one of the major challenges of our time (Boun-four, 2016; Castells, 1996). On a smaller scale, this also affects the world of sport, where the ongoing digitalization and developments in the field of sensor technology have led to a rapid increase in the volume of data. In particular, spatio-temporal position data, which are now available almost everywhere in soccer, harbor huge potential for performance analysis. However, these vast amounts of data have little value in themselves. Rather, the challenge is to develop new methods of data analysis, and to use the resulting structures and correlations in sports data to augment and enhance our knowledge about sport.

The collection, analysis and marketing of data in professional soccer has developed into a business segment with sales in the three-digit million range under the heading of sports analytics (Research and Markets, 2016). There is hardly a club in the world's leading soccer leagues (Premier League, Bundesliga, La Liga, Ligue 1, Serie A, Major Soccer League) today that does not use electronic monitoring of training loads and game data-based analysis of their own teams and their opponents (Miller, 2015; Sands, Kavanaugh, Murray, McNeal & Jemni, 2016; Wright, Atkins, Jones & Todd, 2013). The world's largest data companies like IBM, Intel, SAP, and Microsoft, compete for the best data analytics tools and use soccer as a demonstration domain for the performance of their products. An active scientific community has developed in the academic field. A separate journal for sports analytics, numerous special editions in established journals, textbooks, worldwide conference series and study programs testify to the increasing scientific institutionalization of this research area.

This publication contains six performance analysis studies from the field of professional soccer, which were published in scientific journals in 2016 and 2017. Based on soccer data and mathematical data analysis methods, they illustrate how issues in the context of training and competition can be effectively addressed. This introduction provides an overview to the field of sports analytics, describes the significance of data analysis in the context of professional soccer and organizes the individual publications by thematic context.

1.1 Sports Analytics

In the first instance, *data analytics in sports* (also referred to as s*ports analytics*) is simply the mathematical analysis of sports data. Traditionally, the focus here is on competition data, such as events or actions during the match (e. g. in soccer passes, free kicks, tacklings), spatio-temporal position data from tracking systems, but also information such as match schedules or team formations. These are supplemented by training plans, the results obtained from motor skills tests and biomechanical tests or medical data. We can extend the term 'sports data' to include external data such as contract data, audience numbers or other economic parameters with relevance for club management.

Individual activities in sports analytics can be traced back to the beginning of the 20th century (Fullerton, 1912; Rickey, 1954). In the 70s, the topic became increasingly popular in the academic environment and there was a noticeable increase in the number of scientific publications (Wright, 2009). However, it was not until the book *Moneyball* by Michael Lewis (2003) and the subsequent the film in 2011 that sports analytics gained public attention, and the protagonist of the novel – which is told in reference to a real-life incident – was able to put together a successful sports team for Major League Baseball in the USA with comparatively few financial resources. The key to this success was the skillful combination of players who, on the one hand, were undervalued on the player market, and who had mutually complementary abilities on the other. Financial and sporting decisions were made on the basis of publicly available baseball data and a mathematical evaluation model (Baumer & Zimbalist, 2013).

Originally, the terms analytics or data analytics refer to the search, interpretation and communication of patterns in data (Davenport & Harris, 2007). In the economic context, so-called *business analytics* aims to extract information from company data in order, for example, to provide a basis for the analysis of business performance to support business decisions or improve business processes (Chen, Chiang & Storey, 2012). Similarly, the sports analytics approach consists of searching for patterns in sports-related data and thus optimizing processes and decisions in the sporting field (Alamar, 2013). Although it is not limited to specific sports groups or to a certain level of performance, most of the activities are observable in the top US sports areas of baseball, basketball, ice hockey and American football as well as in soccer since not only are the resources there, but the media and economic interest is great (AT Kearney, 2011).

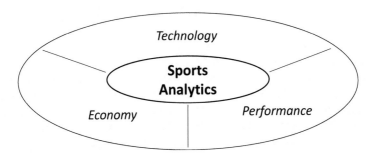

Fig 1.1: Main areas of sports analytics.

Sports analytics is defined for the purposes of this book as

> *the process of searching, interpreting and processing informa-*
> *tion in sports-related performance data using information sys-*
> *tems and mathematical methods of data evaluation with the aim*
> *of achieving competitive advantages.*

These competitive advantages relate, on the one hand, to the sporting sector, in which the performance and efficiency of teams are improved, and on the other hand to the marketing of professional sport to generate economic advantages. In order to differentiate between the sports and economic aspects, the terms *sports performance* analytics and *sports management* analytics are sometimes used in American parlance.

Technological progress has greatly increased the possibilities in both areas. In recent years, the focus has shifted in particular to systems for the automated acquisition of spatio-temporal data (Leser, Baca & Ogris, 2011) and the associated new possibilities offered by data-based tactical analysis. This is also associated with discussion about the appropriate paradigms for analyzing these data (Rein & Memmert, 2016) and the concept of big data. In order to manage the rapidly growing, heterogeneous datasets at clubs and to extract performance-diagnostic relevant information, appropriate information systems are necessary (Shah, Kretzer & Mädche, 2015). In addition to the areas of sports performance and economy, information technology is thus the third main area of sports analytics (Fig. 1.1).

1.2 The Use of Data Analytics in Soccer

The concrete objectives for the analysis of sports data differ depending on the user group. Professional coaching teams in soccer use competition data for tactical match analysis. Here, self-analysis aims to identify weaknesses and reduce them during training. Opponent analyses are mostly used for strategy development in the preparation for competitions. Performance indicators, like presented in Chapters 2 and 3, can be used to obtain quantitative evidence of strengths and weaknesses, or to identify preferred moves or tendencies of action in critical situations (Hughes & Bartlett, 2002; Wright et al., 2013). In reviewing a competition, data can help to check whether the chosen strategy has been implemented, whether it was successful, and which of the assumptions made and conclusions drawn in advance were correct (Cordes, Lamb & Lames, 2012).

Physcial performance indicators from tracking systems, describe the extent and intensity of training and competition (see Chapters 5 and 6). Athletics trainers can use these to individually plan regeneration phases in the microcycle. In addition, the indicators provide evidence of the degree to which the fitness performance requirements are met. Empirical measures can be used to estimate the level of performance in comparison with the collective, and longitudinal data can be used to assess individual performance trends and the effectiveness of training approaches (Mujika & Padilla, 2003).

Medical departments in clubs use training data to minimize the risk of injury by avoiding overloads (Dvorak et al., 2000). Infectious diseases can also be detected at an early stage by means of an unusual ratio of stress and strain parameters, for example running intensity and heart rate. After injuries, longitudinal performance data help assess progress in rehabilitation and provide forecasts of normal performance levels (Blobel, Pfab, Wanner, Haser & Lames, 2017).

In the economic sector, data can be used to support management decisions on transfers of players and player contracts. In view of the rising sums of money for severance payments, salaries and performance bonuses, the challenge lies in identifying talents early on and employing them at favorable contractual conditions (Simmons, 2007; Buraimo, Frick, Hickfang & Simmons, 2015). Better contract design decisions later either generate higher revenues through sales revenue or save high transfer expenses for established players. In soccer today, there are global databases of players and performance data provided by companies like *Opta* or *Wyscout*, with which scouting departments can make a pre-selection of talents or search for undervalued players according to the *Moneyball-hypothesis* (Hakes & Sauer, 2006). If the concept of sports data is extended, activities such as optimizing ticket prices, analyzing fan satisfaction or marketing fan merchandise can be added to this range. At smart stadiums, stadium operators in the USA offer digital services throughout (Panchanathan et al., 2016).

A central application of sports data outside professional clubs is the enrichment of media coverage including game statistics or animations based on position data. In addition, bookmakers use sports data to detect betting fraud (Deutscher, Dimant & Humphreys, 2017), and manufacturers of computer games use it to increase the play value of their products. Worlds largest data companies use sport as a demonstration domain to showcase the performance of their data analytics components (IBM, 2016; Intel, 2017; Microsoft, 2017; SAP, 2015). For leagues and associations, sports data represent an additional marketing opportunity, and, in addition, can also be used for auditing purposes, such as e. g. for the evaluation of rule changes (see Chapter 7).

In addition to being used for commercial purposes, sports data are an attractive object for scientific research (Coleman, 2012). Sports science profits from the enormous data pool from leagues, clubs and sports data companies, which affords the opportunity of investigating academic claims about sports practice, defining performance standards or testing new methodological paradigms (see Chapter 4). Sports data and the issues associated with it are also interesting for computer science, as they are well suited to the development and validation of new approches in computer science due to their degree of complexity. They are neither too simple to be worthwhile investing effort in, nor too complex (limited degrees of freedom through the rules, basic tactical intent) to be unworkable. Sport can thus provide a kind of ëxperimental environment"for computer science, in which human behavior can be observed and investigated in complexity-reduced but nevertheless authentic detail.

1.3 About the Studies in this Book

The studies in this book deal with questions of performance diagnostics in soccer using an empirical paradigm. They are also distinguished by the fact that they use methods of computer science and data sciences in different ways when working on the issues. With this interdisciplinary approach, they fall under the academic discipline of *Sport Informatics* and *Computer Science in Sports* (Link & Lames, 2014), which is represented by the *International Association of Computer Science in Sports* (IACSS). In the tradition of British sports science, this comes under the scientific heading of *Performance Analysis*; the assignment to *Sports Analytics* corresponds to the US-American designation.

Another common feature of the studies is that they are based on data from the official German Bundesliga match database. The *German Professional Soccer League (DFL)* is the only European league to date to have largely centralized the collection and delivery of match data. Since the 2011/12 season, data on all Bundesliga and 2nd Bundesliga matches has been collected by data service providers according to uniform definitions and standards and stored in a central database. Media partners, associations and scientific institutions can access and use this data for their own purposes. Given the enormous volume of data, the potential for performance diagnostics is enormous.

Chapters 2 and 3 illustrate how geometric models may be used to identify tactical structures in spatio-temporal data. Chapter 2, 'Individual Ball Possession', describes an approach for recognizing different types of ball possession which distinguishes between individual and team levels as well as different degrees of ball control. This information is a basic prerequisite for the derivation of higher-value tactical structures – for example, for the danger metric described below in Chapter 3, 'Real Time Quantification of Dangerousity'. It uses the location of possession, density, pressure and ball control to quantify the probability of a goal. Both studies describe the domain-oriented and computational model of the construct, evaluate recognition quality, and give examples of its use in game analysis.

Chapters 4, 5 and 6 include traditional observational studies, which investigate the relationships between independent variables and performance and behavioral parameters in the game. Chapter 4, 'A Topography of Free Kicks', uses geostatistical techniques to describe the dependence of various free kick parameters, such as distance to wall, interruption time, laterality and success of the position of the free kick. Chapter 5, 'Match Importance Affects Player Activity', examines the relationship between the relevance of a soccer match on activity indicators such as running distance, sprints, competitions for the ball, and fouls. For this purpose, a stochastic model is developed that estimates the influence of a match on the team's final rankings. If the probability that a game has an influence on rank is sufficiently high and if it has considerable consequences, such as relegation vs. survival, a game is classified as important, otherwise not. Chapter 6, 'Effect of Ambient Temperature on Pacing Depends on Skill Level', deals with the influence of outdoor temperature on running parameters in soccer. On the basis of data from two Bundesliga and 2nd Bundesliga seasons, the effects of high outside temperatures on running activity are estimated using a regression model. The influence of temperature is also investigated in relation to the level of play.

The study in Chapter 7, 'Vanishing Spray Reduces Extent of Rule Violations', deals with the effects of the introduction of the vanishing spray in the Bundesliga. Free kicks before and after the introduction of the vanishing spray are compared with each other in terms of distance to the wall, rule violations, punishment for infringements of the rules, and success. This is done to parallelize the random samples of the location of the free kick and the number of players in the wall.

The following chapters are versions of the original papers. In order to ensure uniformity in this publication, they have been slightly adjusted. This includes captioning and graphical representation of figures, tables and equations, bibliographies and bibliographies and orthographic conventions. The chapter headings have been slightly adjusted so that some of them do not correspond exactly to the title of the original publication.

2 Individual Ball Possession

Abstract This paper describes models for detecting individual and team ball possession in soccer based on position data. The types of ball possession are classified as Individual Ball Possession (IBC), Individual Ball Action (IBA), Individual Ball Control (IBC), Team Ball Possession (TBP), Team Ball Control (TBC) und Team Playmaking (TPM) according to different starting points and endpoints and the type of ball control involved. The machine learning approach used is able to determine how long the ball spends in the sphere of influence of a player based on the distance between the players and the ball together with their direction of motion, speed and the acceleration of the ball. The degree of ball control exhibited during this phase is classified based on the spatio-temporal configuration of the player controlling the ball, the ball itself and opposing players using a Bayesian network.

The evaluation and application of this approach uses data from 60 matches in the German Bundesliga season 2013/14, including 69,667 IBA intervals. The identification rate was $F = .88$ for IBA and $F = .83$ for IBP, and the classification rate for IBC was $\kappa = .67$. Match analysis showed the following mean values per match: TBP 56:04 $\pm$ 5:12 min, TPM 50:01 $\pm$ 7:05 min and TBC 17:49 $\pm$ 8:13 min. There were 836 $\pm$ 424 IBC intervals per match and their number was significantly reduced by -5.1 % from the 1st to 2nd half. The analysis of ball possession at the player level indicates shortest accumulated IBC times for the central forwards (0:49 $\pm$ 0:43 min) and the longest for goalkeepers (1:38 $\pm$ 0:58 min), central defenders (1:38 $\pm$ 1:09 min) and central midfielders (1:27 $\pm$ 1:08 min). The results could improve performance analysis in soccer, help to detect match events automatically, and allow discernment of higher value tactical structures, which is based on individual ball possession.

2.1 Introduction

Technological innovations of recent years, particularly in the field of tracking systems, are leading to an increasing volume of data in sports. These

© Springer Fachmedien Wiesbaden GmbH, part of Springer Nature 2018

enormous amounts of information present new challenges when it comes to analyzing and interpreting this data. What is required are answers to such questions as how sport clubs can best exploit the possibilities on offer to analyze game tactics, manage training processes and make better transfer decisions, how media companies can use this information to offer better and more innovative match coverage products and how new scientific insights into the nature of sporting phenomena in general and the factors that influence performance can be gained.

In soccer, *Competition Information Providers* (CIPs) and mainstream of sports science traditionally use relatively simple performance indicators, such as shots on goal, number of passes, tackles won, team ball possession, distance covered or heat maps (Castellano, Casamichana & Lago, 2012; Harrop & Nevill, 2014; Vilar, Araújo, Davids & Button, 2012). While some of these standard indicators can be easily generated from the raw data, their usefulness for performance analysis should be regarded with a certain degree of skepticism (Mackenzie & Cushion, 2013; Carling, Wright, Nelson & Bradley, 2014).

In the last years, there has been increased activity in developing more intelligent performance indicators on the part of both the CIPs and the scientific community (an overview is given by Rein and Memmert (2016)). On approach is to model interactions between teams or players based on concepts of dynamic systems theory such as Approximate Entropy or Relative Phase (Lames & McGarry, 2007; Frencken, Poel, Visscher & Lemmink, 2012; Folgado, Duarte, Fernandes, Sampaio & Haddad, 2014). Another group of studies use networks approaches to describe passing behavior and to find patterns which are correlated to success (Gama et al., 2014; Qing, Hengshu, Wei, Zhiyong & Yuan, 2015; Ribeiro, Silva, Duarte, Davids & Garganta, 2017). In addition, Machine Learning is o used to study team tactics. Grunz, Memmert and Perl (2012) for example use self-organizing maps to classify the behavior of small groups of players in set play situations, such as a game opening sequence. Le, Carr, Yue and Lucey (2017) use deep learning algorithms to imitate tactical behavior and estimate how each team might have approached in a given situation.

Another group of emerging approaches to analysis derives from well-known tactical soccer constructs like Control of Space, Availability, Pressing or Dangerousity from spatiotemporal tracking data (Fonseca, Milho, Travassos & Araújo, 2012; Bialkowski et al., 2014; Gudmundsson & Wolle, 2014; Spearman, Basye, Dick, Hotovy & Pop, 2017; Link, Lang & Seidenschwarz, 2016). These construct can be organized using a hierarchy consisting of actions and different types of situations (Beetz et al., 2009), which can be

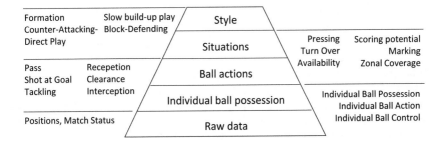

Fig 2.1: Hierarchy of soccer constructs. Individual Ball Possession and its subtypes are a fundamental prerequisite for being able to discern ball actions. These actions are again the base for identifying situations. The typical behavior in these situations can be aggregated to playing styles.

aggregated to playing styles as shown in Fig. 2.1. A fundamental prerequisite when discerning such constructs is to know, which player has the ball. This information is typically not included in the raw data provided by CIPs. The reason is, that ball possession data is collected by human data loggers concurrent with the game in real time, and it would be too expensive to manually record data on an individual player basis. To our knowledge, there are no studies, which deal with the problem of identifying ball possession on an individual level.

Furthermore, Individual Ball Possession is not only important as an auxiliary concept. In soccer's performance analysis, team ball possession is the most commonly investigated performance indicator (Mackenzie & Cushion, 2013). Its relevance is easy to understand, since having control of the ball is a fundamental prerequisite for being able to invade the opposing team's third of the pitch and score goals (Bate, 1988). As a consequence, successful teams not only have a greater share of ball possession (Hughes & Franks, 2005; Grant, Williams, Reilly & Borrie, 1999) but the periods of ball possession are longer too (Jones, James & Mellalieu, 2004). On the other hand, having more ball possession is not of itself a criterion for success. As such, teams tend to exhibit less ball possession in won games than they do in lost games – a phenomenon that may be explained by a change in tactics depending on if they are in the lead or behind (Jones et al., 2004; Lago-Peñas & Dellal, 2010). Ball possession is not so much a causal variable but rather the consequence of a process of interaction that is determined by a number of contextual factors, such as the venue, the quality of the oppon-ent, the tactical configuration and the current score (Bloomfield, Polman & O'Donoghue, 2005; Lago & Martín, 2007; Pratas, Volossovitch & Ferreira, 2012; Collet, 2013).

This paper is the first to describe how ball possession data can be collected and evaluated not only at the team level, but at an individual level. There is one paper by Kang, Hwang and Li (2006) that determines ball possession using an approach similar to that presented in this paper, but the methods differ in their basic objective, however, as they examine more general as-pects of player trajectories using simulated data. In the following sections, we first define various types of ball possession independent of operational considerations. Computational models are developed for calculating these variants and validated using manually collected reference data. Lastly, we present and discuss some applications for the presented models. The results can be used by other scientific workgroups who need information on Indi-vidual Ball Possession to analyze tactics and also by coaches and analysts to improve their capabilities in performance analysis.

In the following sections, we first define various types of ball possession inde-pendent of operational considerations. Computational models are developed for calculating these variants and validated using manually collected refer-ence data. Lastly, we present and discuss some applications for the presen-ted models. The results can be used by other scientific workgroups who need information on Individual Ball Possession to analyze tactics, and by coaches and analysts to improve their capabilities in performance analysis.

2.2 Ball Possession Types

In the following, a system of terms is introduced that defines the phe-nomenon of ball possession and differentiates its various types (Fig. 2.2). In our approach, only time intervals during which the ball is in play are considered when determining ball possession. When the ball is in play, one of the two teams always has team ball possession and one of the play-ers always has individual ball possession. When individual ball possession switches between two players, it is assumed that the pass reflects the tac-tical intent of the first player. This time interval is thus classified as ball possession for the first player. No individual player and neither team is assigned ball possession when the ball is not in play. This approach differs from that taken by some CIPs, who ascribe these time intervals to one or

other of the teams (either the team that had possession of the ball up to
that moment or the team that is in possession of the ball thereafter).
One detail of the definition concerns the question of when ball possession
begins. In order to simplify the data collection process, CIPs often set
this time to the moment of first contact with the ball. In the following,
however, the ball is considered to be in a player's possession once it enters
that player's action space (for example, a slow ball that a player is close to
even though he has not yet made contact with it). Ball possession is thus
defined as follows:

1. *Individual Ball Possession* (IBP) begins the moment a player is able to
   perform an action with the ball (following an IBP of another player or a
   game interruption). It ends the moment IBP for another player begins.

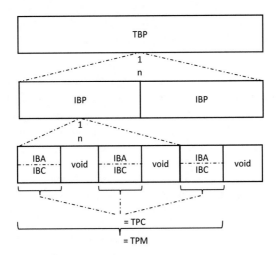

Fig 2.2: Relationships between the different types of ball possession. Team ball
possession (TBP) consists of an unbroken sequence of individual ball
possession (IBP) phases. IBP is composed of one or more phases in
which an action can be performed with the ball. These are designated
individual ball actions (IBAs). It is possible for IBAs to be separated
by void phases during which no ball control exists (e.g. while the ball
is in the air). An IBC is an IBA where ball control is present. Team
ball control (TBC) is the union of all TBC phases. Team play making
(TPM) corresponds to TBC plus the intermediate void phases.

2. *Team Ball Possession* (TBP) begins the moment IBP for one of the team's players begins (following IBP of a player on the opposing team or a game interruption). It ends with the first IBP for one of the opposing team's players following this.

The time interval in which a player can perform an action with the ball is then separated out from this simple definition of ball possession. This interval no longer includes the time from the ball being passed on until the start of IBP of the next player. It is important to make this distinction in view of the fact that the configuration on the pitch during this time interval determines the tactical options available and their chances of success. The time interval during which no influence can be exerted on the ball is unimportant here.

3. *Individual Ball Action* (IBA) of a player begins the moment this player is able to perform an action with the ball and had no IBA prior to this. It ends the moment the player is unable to perform any further action with the ball.

A further internal distinction within the IBA can be made based on the level of ball control exhibited. What matters here is whether a player has the ball sufficiently under control that he can consciously chose between several play options. An example where this is not the case is when a player attempts to deliver the ball to a specific area of the pitch under extreme pressure (or 'ping-pong' sequences in the midfield or the ball being headed on following a cross). This distinction is made because it is only possible to draw reliable conclusions about team or individual tactics during game analysis if the ball is fully under control.

4. *Individual Ball Control* (IBC) for a player begins when IBA for this player begins and he is able to decide between several play options during the IBA. It ends the moment this particular IBA of the player ends.

Two further constructs are defined at the team level. Team ball control is the union of all IBCs, i.e. the period of time during which any player in the team has control of the ball. This very narrow understanding of team ball control is supplemented by the 'Team Playmaking' construct. This includes the periods of time during which the ball is being passed between the players of a particular team as well. It is equivalent to the common understanding of team ball control prevalent for coaches today.

5. *Team Ball Control* (TBC) for a team begins when IBC for one of this team's players begins and ends as soon as the IBC of this player ends.

6. *Team Playmaking* (TPM) for a team begins when IBC for one of this team's players begins and the team has had no IBC immediately prior to this. It ends with the last IBC before the next IBA of a player on the opposing team.

2.3 Detection of Ball Possession

Automatic detection of ball possession involves a multi-step process (Fig. 2.3). The first step involves pre-processing raw data provided by the CIPs in order to reduce the number of unrealistic position jumps and noise in the player and ball coordinates. The core of the procedure involves detecting the moments where the IBPs and IBAs start and the IBAs finish. As shown in the following sections, these time codes can be used to reconstruct all of the required time intervals. After that, the level of ball control within the IBAs is estimated using a Bayesian network. The following sections describe the individual steps of the process in detail.

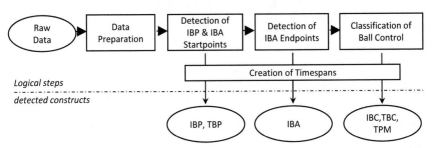

Fig 2.3: Logical procedural steps for detecting different types of ball possession. Detection is based on identifying the starting and endpoints of IBP and IBA.

2.3.1 Data Preparation

The raw data exists as a sequence of data frames with a frame rate of 25 Hz. This paper only considers the x/y coordinates of the players and the ball, and the running flag (which measures whether the ball is in play versus out of play). Some CIPs also provide height information as well as a team possession flag, but these data are not relevant for determining the various types of ball possession that is the focus of this study.

The first step involves identifying unusable sequences in the raw data and excluding them from further processing. First, jumps in the ball's trajectory are detected by applying distance tests on temporally adjacent data points. Very short sections ($< 1\,$s) are excluded from further processing. The Rauch-Tung-Striebel (RTS) method (Särkkä, 2013) is used to smooth the trajectories of the ball and the players in the remaining sequences. The smoothed data are then used to calculate distances, velocities and accelerations. In the following, for any time code t, the position of the ith player in the raw data is denoted by $x_t^i \in R^2$ and that of the ball by $b_t \in R^2$. After applying the stochatical RTS-smoother, the positions are given by $\underline{x}_t^i$ and $\underline{b}_t$ the x/y-speeds by $\underline{\dot{x}}_t^i$ and $\underline{\dot{b}}_t$ and the x/y-acceleration of the ball by $\underline{\ddot{b}}_t$. The distance $d(a, b)$ between two positions corresponds to the Euclidean distance.

2.3.2 Detection of IBP & IBA Starting Point

The IBP or IBA starting point is the moment when a player starts to interact with the ball. We state this as soon as the distance between the player and the ball falls below a threshold value and that player is nearest to the ball. Put formally, this means that both the formula

$$d(\underline{x}_t^i, \underline{b}_t) < T_P \tag{Eq. 2.1}$$

Becomes true at time t and the distance on the left-hand side of the equation is not smaller for another player id $j \neq i$. The value T_P determines the threshold value for a player being able to physically interact with the ball and has to be trained a priori. We refer to this method as *naive physical* (Fig. 2.4, left).

As stated earlier, the ball position is provided without a z-coordinate. If the ball passes above a player, this could easily lead to possession being incorrectly attributed. We obviate this problem by applying not only a 2d separation threshold, but also by testing to check if a player has interacted with the ball. We use the local maxima ball accelerations together with narrow tolerance to filter out noise. A player only gains ball possession if the requirements for *naive physical* are fulfilled (using another threshold value T_K) and a local maxima with a minimum acceleration of $4\,\mathrm{ms}^{-2}$ can be detected. We refer to this method as *kick detection* (Fig. 2.4, right).

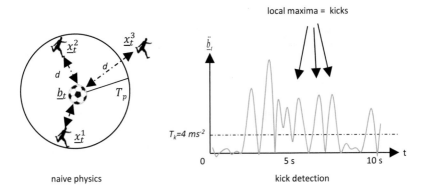

Fig 2.4: Detection of IBA starting point. According to the naive physics model, ball possession exists at time t when the separation d between player $(\underline{x}_t^i)$ and Ball $(\underline{b}_t)$ is less than the threshold T_p. Ball possession is attributed to the player closest to the ball. With kick detection (right), the acceleration of the ball $\underline{\ddot{b}}_t$ must simultaneously show a local maximum of at least $4\,\mathrm{m\,s^{-2}}$.

2.3.3 Detection of IBA Endpoint

While the IBP endpoints are known once the IBA starting points have been identified (a player's ball possession ends when there is an interruption of play or when another player has IBP), the IBA endpoints must be specifically identified. There are three possible ways an IBA for a player can end: 1) the game is interrupted, 2) another player gains IBA or 3) the player is no longer able to interact with the ball. Whereas the first two cases are trivial to detect (by checking the running flag or by detecting a new IBA starting point for a different player) the last one requires special treatment. This involves checking whether a player will be still able to interact with the ball within a certain time span. This is true if, for example, a player kicks the ball a few meters ahead of himself while dribbling, but not after making a pass or taking a shot at goal.

Two approaches are presented here. Both make use of the current positions and velocities of the ball or the ball and player to give an estimate of their future locations. Beginning with a constant prediction interval of 1 s, we can define a future ball position as

$$b_t^+ := \underline{b}_t + \underline{\dot{b}}_t \qquad \text{(Eq. 2.2)}$$

(to simplify the equations, we have omitted the underscore for new variables). The individual ball action interval is now defined to end at the first moment t after it began in which the statement

$$d(\underline{x}_t^i, \underline{b}_t^+) > T_A \tag{Eq. 2.3}$$

is true for future ball position $\underline{b}_t^+$. Again, the threshold value T_A has to be trained a priori. In other words, as long as it is possible for the player currently in possession of the ball to control the ball in the near future, he will retain ball possession. This *ball prediction* model (Fig. 2.5, left) can be refined still further by incorporating a prediction about the player's future position

$$x_t^{i+} := \underline{x}_t^i + \underline{\dot{x}}_t^i \tag{Eq. 2.4}$$

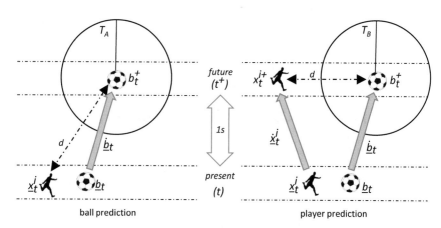

ball prediction   player prediction

Fig 2.5: Detection of IBA endpoint. An endpoint exists as soon as the player is unable to interact with the ball for a period of one second. This is the case after a shot on goal or a pass, for example, but not when the player is dribbling the ball stated formally, when using the ball prediction method, the endpoint t of an IBA phase is arrived at when the distance d between ball position (b_t^+) and player position $(\underline{x}_t^i)$ is no longer below the threshold value (T_A) for a period of one second (t^+) at the ball's current speed and direction $(\underline{\dot{b}}_t)$. The player's position is assumed to be constant. The player prediction method additionally takes into account the player's future position x_t^{i+} based on his direction and speed $(\underline{\dot{x}}_t^i)$.

and changing the inequality to

$$d(x_t^{i+}, b_t^+) > T_B \qquad \text{(Eq. 2.5)}$$

We refer to this as *player prediction* (see Fig. 2.5, right). In contrast to the previous method, the distance check now incorporates two predictions of future position.

2.3.4 Classification of IBC

Once the IBA starting and endpoints have been detected, the IBA intervals can also be derived. The central question is therefore which of the intervals represents a segment that features ball control. This obviously depends upon a great number of factors that may not be possible to ascertain from positional information alone. For this reason, we decided to use a Bayesian network to classify ball control based on the following features (referring to a particular IBA interval): IBA duration, average ball velocity and acceleration, variance of ball velocity and acceleration, average distance between the ball and the player in possession of the ball and the number of opposing players within certain distances (0.5 m, 1 m, ..., 5 m). Training of the net was performed using the K2 algorithm (Cooper & Herskovits, 1992).

2.3.5 Creation of Time Intervals

The classified moments can now be used to unambiguously calculate both the IBPs and the IBAs. Status transitions in ball possession are also determined technically during interruptions in play. Phases when the ball is not in play are not excluded until the time intervals are calculated. This allows for greater tolerance of imprecision when it comes to the running flag.

In this sense, an IBP interval is derived from a determined starting point. The next detected IBA starting point or an interruption in play functions as the endpoint. IBA intervals are calculated from pairs of starting and endpoints (while taking into account interruptions in play). The intervals are each tagged with the player in control of the ball. The team-specific construct TBP is the union of all IBP intervals for a particular team. The TBC is the union of the IBCs of all the players in a team; the TPM, on the other hand, is, according to the model, the result of two new sequences of intervals (one for each team). For each interval in a sequence, the starting point is the beginning of the first IBC of a team following a stoppage or the opposing team being in possession of the ball. The end of such a sequence is the final IBC of the associated TBP phase.

2.3.6 Limitations

The procedure does not consider all special cases that might occur during a match. Firstly, for being able to perform an action with the ball, it is not obligatory to touch it. In the case that a player runs beside the ball for a quite long period without kicking it, the IBA interval would start in the moment of the first touch, which is maybe too late. On the other side, it is possible to touch the ball without reaching the acceleration threshold. This could especially happen, when the contact does not change the ball direction substantially. In other cases, it could be wrong to assign ball possession to the nearest player, e.g. when this player is standing with his back to the ball. In addition, the assessment of ball control depends on the individual skills of a player and is beyond the scope of this paper.

2.4 Training & Evaluation

The training and evaluation phase had two sub-goals: Firstly, training was performed to determine the model parameters. Secondly, the quality of IBP, IBA, IBC detection was examined. An evaluation of the team-specific constructs TBP and TBC was not necessary, since they are derived entirely from the IBPs and IBCs.

2.4.1 Training of Threshold Values & Bayesian Network

Five matches of the sample described in Section 2.5 served as the test sample for training and evaluation. An annotation with reference data (ground truth), which was manually logged by a trained, independent observer post-match, formed the basis for the evaluation. The observer had a panoramic video image of the game at his disposal together with a graphic overlay of positions, frame-by-frame playback control and as much time as needed. A total of 6,976 IBA phases with starting and endpoints were identified. 6,340 of these included ball control.

The training to establish the threshold values T_P, T_K, T_A, and T_B that appear in the model was performed using the reference data. A full grid search was performed in order to determine the global optimum for each

Tab. 2.1: Quality of identification of different individual ball possession intervals. IBP^N = Starting point: Naive physical, Endpoint: /; IBP^K = Starting point: Kick detection, Endpoint: /; IBA^B = Starting point: Kick detection, End point: Ball prediction; IBA^B = Starting point: Kick detection, End point: Player prediction.

	Recall	Precision	F_1-Score	Timeline
IBP^N	.90	.68	.77	.82
IBP^K	.80	.86	.83	.87
IBA^B	.86	.91	.88	.92
IBA^P	.84	.92	.87	.91

of the model's threshold values (Powell, 1988). The Bayesian network described in Section 2.3.4 was trained using data from the second half of the matches and evaluated using data from the first half. The test data contained 1,674 IBA intervals of which 107 involved no IBC. The aim was to identify these. The training data exhibited a similar ratio with 1,583 to 122.

2.4.2 Quality of IBA & IBP Detection

The verification of the running flags and detection of IBA starting and endpoints was performed by comparing them with the respective status change in the ground truth. A change detected in the data is interpreted as a *true positive* (TP) if a change is also present in the ground truth within a 0.6 s interval and – in the case of ball possession – the associated player is correct. A *false negative* (FN) occurs when there is a change in the ground truth that is not detected in the data. A change is classified as a *false positive* (FP) if there is a change detected in the data that is not present in the ground truth.

The metrics Precision, Recall and F_1-Score were used to assess the quality of the recognition system, as is usual in machine learning (Baeza-Yates & Ribeiro-Neto, 1999). A Timeline quantity is also introduced. This involves the superimposition of the time axes for the ground truth and the IBP or IBA data with respect to ball possession and comparing each individual point in time. A point in time is adjudicated correct if the same ball possession was present in the ground truth within the specified time window of 0.6 s. The ratio of correct time codes to the number of time codes within the net playing time is recorded as a Timeline quantity.

Tab. 2.1 shows the identification rates for IBP and IBA. The results suggest that both *kick detection* and *ball prediction* are the most promising models. Although there is greater inaccuracy with regard to ball position, prediction of a player's position and the knowledge of the player's speed do not lead to better concordance as regards classifying ball possession, at least not according to the selected model. In terms of IBA (ball prediction), the data have a 5.9 with respect to timeline.

When analyzing video sequences for the FP and FN intervals we found, that many errors can be attributed to tracking losses and running flag inaccuracies. These are fundamental problems that similarly affect every CIP and every game. Some CIPs are already offering to manually edit the tracking data afterwards in order to increase the quality of the data. Nevertheless, methods of analyzing play structures must take these fundamental flaws in the raw data into account and implement appropriate error detection and correction procedures.

On the other hand, one can safely conclude that the quality of tracking has arrived at a level allowing for tactical structures to be reliably detected. The rate of detection of individual ball possession should be good enough for answering many questions regarding performance. As tracking quality will surely increase in the next years due to the technological progress of CIPs, detection will also become more stable. However, current quality clearly exceeds the accuracy of manual acquisition by human game loggers (52 %). It is easier for a computer to observe the tolerance of 0.6 s than it is for a human. In addition, manual data acquisition involves a great deal of effort and is only performed at the team level.

2.4.3 Quality of IBC Classification

The quality of recognition of IBCs for a given IBA interval was also determined by a comparison with the ground truth. All IBA sequences from the second half (n = 1,674) were used as the data set here. Tab. 2.2 shows the results as a confusion matrix.

97.0 % of intervals involving ball control were correctly classified. If no ball control was present, however, only 50 % of the intervals could be correctly classified. Overall, 96.7 % of the intervals are correctly attributed. The degree of consistency according to Cohen is $\kappa = .67$. However, with only 122 non-IBC intervals, the training set for differentiating between IBAs and IBCs is very small. Moreover, the inter-rater reliability test between two human observers on a subset of the intervals (n = 98) did not show complete consistency either ($\kappa = .72$). This indicates that it may not be possible to

Tab. 2.2: Results of IBC classification. 96.7 % of IBA intervals were correctly classified using the Bayesian network.

		Ground Truth	
		IBC	No IBC
Bayesian Network	IBC	1,535	23
	No IBC	32	84

fully objectify the ball control construct. This is quite typical for non-trivial tactical concepts and has also been reported e.g. for the *Dangerousity* metric (Link et al., 2016).

2.5 Game Analysis

As a first application of the method, we evaluated new performance metrics based on the different types of individual ball possession. Our sample comprises 60 matches during the 2012/13 seasons of the German Bundesliga. The positional data were collected during the match using an optical tracking system (TRACAB). Kick detection was used to calculate the IBP and TBP, IBA, IBC and TBC intervals were determined using ball prediction without player prediction. Overall, we collected 69,667 IBA intervals, including 53,354 IBCs.

2.5.1 Team Based Metrics

Tab. 2.3 shows the evaluation of gross game time, net game time (excluding game interruptions), TBP, TMP and TPC. TBP is equal to the net playing time by definition. Since not every TBP involves ball control (e.g. in the case of ping-pong sequences in the midfield) and final phases that do not exhibit control are not counted in this time interval either, the TPM is lower than TBP. The accumulated time of TBC intervals is even lower, because the time from the ball being passed on until the start of IBC of the next player is excluded. The ball possession variables on team level are moderated correlated with net playing time (TPM (min), r = .40; TBC (min), r = .38; TBC (n), r = .24).

A paired t-test shows that net game duration was reduced by -2.4 % (effect size d = 0.2) from 1st to 2nd half (t = 2.0, p < .05) although gross game duration in 2nd half was $+4.1$ % (d = 1.8) higher compared to 1st half

Tab. 2.3: Team based metrics. Playing time (PT), TPB, TPM, TBC according to game section and team status.

	Σ	1st half	2nd half	Home	Away
PT gross (min)	93:32 ± 1:36	45:50 ± 0:45	47:43 ± 1:22	28:28 ± 6:06	27:35 ± 6:35
PT net (min)	56:04 ± 5:12	28:22 ± 3:19	27:43 ± 2:38	25:32 ± 5:45	24:39 ± 5:09
TBP (min)	56:04 ± 5:12	14:29 ± 3:23	13:53 ± 3:55	9:14 ± 5:12	8:34 ± 4:34
TPM (min)	50:01 ± 7:05	12:54 ± 2:51	12:27 ± 2:43	429 ± 243	407 ± 222
TBC (min)	17:49 ± 8:13	4:44 ± 2:42	4:25 ± 2:28	1:39 ± 0:56	1:32 ± 0:49
TBC (n)	836 ± 424	208 ± 123	209 ± 116	76 ± 43	73 ± 40
TBC (min$^{-10\,\text{min}}$)	3:11 ± 1:28	0:51 ± 0:29	0:47 ± 0:26	28:28 ± 6:06	27:35 ± 6:35
TBC (n$^{-10\,\text{min}}$)	149 ± 77	39 ± 22	37 ± 21	25:32 ± 5:45	24:39 ± 5:09

(t = 2.0, p < .001). A similar significant reduction can be observed for TPC(n) and TPC (min). This might be explained by the tactical use of game interruptions in soccer. Siegle and Lames (2012) found out that goal kicks or free kicks e.g. goal kicks of the leading team take longer towards the end of the match. Also, there might be a tendency of home teams to have more ball possession, which was also reported by Lago-Peñas and Dellal (2010), but the differences do not reach significance level ($\alpha = .05$) in this sample.

In order to exclude the effect of different net playing time on ball possession variables, we also calculated the number and time of TBC per 10 minutes of net playing time. TPC $(n^{-10\,min})$ was reduced by -5.1 % (d = 0.1) from 1st to 2nd half (t = 2.1, p < .001). The same can be observed for TPC $(n^{-10\,min})$. This finding suggests that the reduction of individual ball possession is not only attributed to less net playing time, but also to a change of game dynamics. This is in line with the findings of Harper, West, Stevenson and Russell (2014), who reported a reduced number of passes in the extra time of matches and other studies, that found a decline in running activity from 1st to 2nd half (Carling & Dupont, 2011; Di Salvo, Pigozzi, González-Haro, Laughlin & De Witt, 2013). One explanation for this could be fatigue, caused by physiological factors, or a conscious or subconscious pacing strategy (Paul, Bradley & Nassis, 2015). Whatever the case, our data at least suggest that these factors also influence individual ball possession.

2.5.2 Player Based Metrics

Tab. 2.4 shows the values for IBP, IBA and IBC at player level. Data from players, who played less than 45 min in a match, was excluded. In general, one can see that IBP (n) is somewhat smaller than IBA (n), because IBP can include several phases of IBA (if the ball is briefly outside the player's area of action, but subsequently regains control of the ball without any intervening IBA of an opposing player). IBP (n) data corresponds to the 'ball contacts' construct that match data providers collect manually using loggers. IBC is lower than IBA, as is also seen at the team level, because not every ball contact involves ball control.

The average IBC duration, i.e. the mean time interval in which a player in possession of the ball can make and execute a tactical decision, was 1:21 ± 1:13 min for players who played for the entire gross match duration. A one-way ANOVA indicated significant differences between the tactical positions for the count of IBP (F = 19.8, p < .01), IBA (F = 19.1, p < .01) and IBC (F = 15.5, p < .01). The highest number of ball possession intervals

Tab. 2.4: Player based metrics based on individual ball possession according to playing position. $^{-10\,min}$ indicates occurrences / time per 10 min net playing time. IBC (sn^{-1}) represents mean time per interval. Playing positions are GK = Goalkeeper, FB = full back, CD = central defender, WF = wide midfielder, CM = central midfielder, CAM = central attacking midfielder, WI = Winger, CF = central forward and ALL = all positions.

Position	IBP $(n^{-10\,min})$	IBA $(n^{-10\,min})$	IBC $(n^{-10\,min})$	IBC (min)	IBP $(min^{-10\,min})$	IBC (sn^{-1})
GK	9.8 ± 4.3	10.1 ± 4.2	8.8 ± 4.8	1:38 ± 0:58	17.3 ± 10.1	2.1 ± 0.8
FB	14.8 ± 6.8	15.0 ± 6.7	11.7 ± 7.7	1:18 ± 0:56	13.8 ± 10.0	1.2 ± 0.4
CD	15.5 ± 7.8	15.9 ± 7.8	12.8 ± 8.3	1:38 ± 1:09	17.1 ± 12.2	1.3 ± 0.4
WM	13.7 ± 7.2	13.8 ± 7.2	10.2 ± 7.7	1:17 ± 1:08	13.6 ± 12.1	1.2 ± 0.6
CM	16.5 ± 8.6	16.8 ± 8.5	12.8 ± 9.0	1:27 ± 1:08	15.2 ± 12.0	1.2 ± 0.4
CAM	14.0 ± 7.3	14.3 ± 7.5	10.2 ± 8.0	1:17 ± 0:58	13.6 ± 10.2	1.1 ± 0.6
WI	11.9 ± 7.7	12.2 ± 7.8	7.9 ± 7.3	1:04 ± 0:54	11.4 ± 9.7	1.1 ± 0.7
CF	10.1 ± 6.4	10.6 ± 6.4	6.6 ± 6.0	0:49 ± 0:43	8.7 ± 7.6	0.9 ± 0.6
ALL	13.3 ± 7.2	13.6 ± 7.2	10.1 ± 8.1	1:21 ± 1:13	13.8 ± 12.6	1.2 ± 1.0

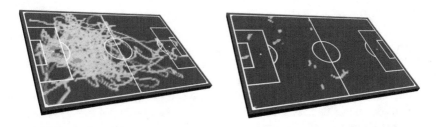

Fig 2.6: Heatmaps of a central forward. Visualization based on all positions (left) compared to positions during IBC (right).

were found for central midfielders (CM), followed by central defenders (CD). Goalkeepers (GK) and central forwards (CF) showed the lowest numbers of ball possession intervals. This is easy to explain since CM and CD are most responsible for play making. Regarding their relative position in the team formation, CF are the closest to the opponent's goal which means, that they are most difficult to reach by passing. This is supported by Fig. 2.6, which shows a 'traditional' heatmap, including all positions during the match (a) for a CF in contrast to a heatmap based only on the time periods where the player owned IBC (b). Most of the IBC intervals occur in the opponent's half.

It may also be possible to draw conclusions about the playing characteristics of a playing position by comparing the duration of the ball possession intervals. Although there is a strong correlation between IBC (n) and IBC (min) (r = .85), significant differences in the average lengths of the IBC phases can be detected (F = 53.5, p < .01). For example, CFs have the shortest average ball contact time ($0.9 \pm 0.6\,$s), which can be explained by what happens when they had possession of the ball (most of the time being immediately tackled by an opposing player). In contrast, GKs had the longest IBC intervals and the highest proportion of IBA intervals with control. The reasons might be that they are allowed to catch the ball and then are not open to attack.

2.6 Conclusion

The evaluation suggests that the method is suited for determining individual ball possession in soccer. It can be used for a wide variety of potential applications. Firstly, it allows quantification of the amount of time a player in possession of the ball spends in different areas of the pitch or the distribution of ball possession times. Secondly, information about basic events, such as passes, tackles, or shots on goal, can be deduced directly from the individual ball possession data. These events are logged by CIPs as default, but due to the manual data collection process the event lists are not always complete and the events' time stamps are sometimes imprecise. Automatic detection methods can thus help to ensure the quality of match data and potentially reduce the loggers' workload. Thirdly, being able to detect ball possession is a fundamental prerequisite for discerning higher value tactical structures, such as the ability for players to receive a pass, pressing strategies or marking tactics. Finally, the ability to recognize ball possession types holds considerable potential for improving the quality of match analysis in professional soccer.

Acknowledgments   This chapter is taken from Link, D. & Hoernig, M. (2017). Individual ball possession in soccer. *PLOS ONE*, *7*(5). doi:10.1371 /journal.pone.0179953.

3 Real Time Quantification of Dangerousity

Abstract   This study describes an approach to quantification of attacking performance in soccer. Our procedure determines a quantitative representation of the probability of a goal being scored for every point in time at which a player is in possession of the ball – we refer to this as dangerousity. The calculation is based on the spatial constellation of the player and the ball, and comprises four components: (1) Zone describes the danger of a goal being scored from the position of the player on the ball, (2) Control stands for the extent to which the player can implement his tactical intention on the basis of the ball dynamics, (3) Pressure represents the possibility that the defending team prevent the player from completing an action with the ball and (4) Density is the chance of being able to defend the ball after the action. Other metrics can be derived from dangerousity by means of which questions relating to analysis of the play can be answered. Performance quantifies the number and quality of the attacks by a team over a period of time, while Dominance describes the difference in performance between teams. The evaluation uses the correlation between probability of winning the match (derived from betting odds) and performance indicators, and indicates that among Goal difference (r = .55), difference in Shots on Goal (r = .58), difference in Passing Accuracy (r = .56), Tackling Rate (r = .24) Ball Possession (r = .71) and Dominance (r = .82), the latter makes the largest contribution to explaining the skill of teams. We use these metrics to analyze individual actions in a match, to describe passages of play, and to characterize the performance and efficiency of teams over the season. For future studies, they provide a criterion that does not depend on chance or results to investigate the influence of central events in a match, various playing systems or tactical group concepts on success.

3.1 Introduction

The availability of virtually all-encompassing positional data in professional soccer presents new challenges for the way in which that data is analyzed and interpreted. They relate equally to analysis of games in clubs, product design for reporting in the mass media, and new analytical procedures for addressing academic questions. A significant factor in this context comprises the description of the technical-tactical aspects of the events of a match by means of performance indicators (Hughes & Bartlett, 2002). Although traditional indicators, such as shots on goal, number of passes, tackle rates, team ball possession and distances covered are widely used, their significance for performance is open to critical question (Carling et al., 2014; Mackenzie & Cushion, 2013). The key task for data science and sports science is to derive intelligent indicators from raw data that describe relevant components of the game appropriately.

Recent years have seen an increasing number of publications that report successes in identifying tactical structures. Grunz et al. (2012) use self-organizing maps to classify the behavior of small groups of players in set play situations, such as a game opening sequence. Bialkowski et al. (2013) present a method that can adaptively assign roles played by individual players. Similarly, playing styles can be described through the spatial distribution of plays (Lucey, Oliver, Carr, Roth & Matthews, 2013) or the characterization of ball possession phases through gains in territory, the number of passes or the speed of play (Kempe, Vogelbein, Memmert & Nopp, 2014). From retrospective analysis of goals and shots on goal, promising spatial constellations can be classified (Wei, Sha, Lucey, Morgan & Sridharan, 2013) or metrics of network analysis can be used to describe the proportion of individual players involved in the team's success (Duarte, Araújo, Correia & Davids, 2012; Duch, Waitzman & Amaral, 2010).

This paper suggests a solution to a question that has largely been unresolved to date, namely: How can success in soccer be quantified? Until now, there has been no convincing procedure available by means of which the value of a piece of dribbling can be compared with a pass, or various passing options compared with one another. If a coach wants to know whether a change in defensive midfield has led to greater stability in defense, he has not so far had any quantitative criterion that would allow such an assessment. Conclusions about the general success of tactical measures against an opponent who is sitting deep, for example, also require a yardstick by which 'more successful' can be measured.

When we use the term 'success' in the sense of performance analytics, we are not referring to the outcome of the game. Goals are scored only rarely in soccer, and can come about through an individual moment of loss of concentration, while a very dominant team might simply be unlucky sometimes. Rather, in order to answer the question posed above, a criterion is required that allows an evaluation of the extent to which tactical objectives were achieved. We believe that creation of situations in which there is a danger of a goal being scored or the prevention of such situations for the opponent, should be the central criterion in characterizing tactical success, or 'performance' in soccer. Shots on goal may be a better criterion than goals in this context, although they may arise in situations that are not dangerous or a player may be prevented from shooting just a few meters in front of goal. In the semi-final of the 2014 World Cup, for example, Brazil had more shots on goal than Germany (18 vs. 14) (FIFA, 2014), but hardly any observer would doubt Germany's superiority in that match (result 1:7). Therefore, our approach to describing success does not use events but a quantitative representation of the probability of a goal, which we describe as *Dangerousity*. We calculate this value for every moment during which a player is on the ball. Dangerousity is related to the construct 'scoring opportunity', but its quality is not evaluated by guesswork but by a defined process using an algorithm.

Additionally, we derive other metrics from dangerousity by means of which questions relating to analysis of the play can be answered. *Action Value* represents the extent to which the player can make a situation more dangerous through his possession of the ball. *Performance* quantifies the number and quality of the attacks by a team over a period of time, while *Dominance* describes the difference in performance between teams.

Our modelling procedure follows the paradigms of rationalism and deduction rather than empiricism and induction (Markie, 2015). In other words: our starting point is our understanding of soccer. We believe that dangerousity is mostly determined by four factors: (1) the position, so called *Zone* of the ball, (2) the degree of *Ball Control*, (3) the *Pressure* that is put on the player by the opponent and (4) the *Density* of opponent players in front of the goal. While there are other factors, we suggest that these four components are the key indicators. To operationalize these indicators, we use mathematical functions which take spatiotemporal data as their input. We choose their functional form in a way that fulfils our analytical understanding of the game (e.g. a defender behind an attacker creates less pressure compared to a defender in front of the attacker).

To date there have been two similar approaches in basketball (Cervone, DAmour, Bornn & Goldsberry, 2016) and in soccer (Lucey, Bialkowski, Monfort, Carr & Matthews, 2015). In these, the probabilities of success are also described continuously by means of a so-called *expected position value* (EPV) or an *expected goal value* (EGV). The approach in soccer determines this value on the basis of position, distance from defenders and the match context (e.g. open play, counter-attack). In contrast to our procedure, which takes account of all match situations at a distance of less than 34 m from the opponent's goal line, in EGV calculations, only the last 10 seconds before shots on goal are considered. The procedure is also based on a less explicit modelling of the individual components that make up the danger of a goal being scored. It is also worth mentioning that the company *ProZone* markets a construct of dangerousity (Komar, 2015), but details of the way it operates have not been published to academic standards.

The aim of this paper is threefold: Firstly, it shows, how dangerousity and derived metrics are quantified. Some details of the specification are left to one side; instead, the focus is on the underlying ideas and relationships. Secondly, an evaluation of the quality of the quantification is carried out, together with the quantitative evidence for the construct's relevance to performance. In the view of the authors, these components of validation, in particular, are not sufficiently documented by many *competition information providers* (CIP) even though they represent a central component in the development of performance indicators in sports. To our knowledge, this paper is the first to use the correlation with betting odds as a criterion for the relevance of performance indicators. Thirdly, the paper shows examples of how the metrics developed can be used to answer questions relating to analysis of a match with different time horizons.

3.2 Quantifying Dangerousity

3.2.1 Dangerousity

Dangerousity (DA) is present for every moment in which a player is in possession of the ball - and can therefore complete an action with the ball. We refer to this time span as *Individual Ball Action* (IBA) and to the player concerned as the *IBA-player*. IBA exists as soon as the distance between the player and the ball falls below a threshold and the ball is then touched. IBA ends when the ball is out of the player's range once again. The procedure is described in detail in Hoernig, Link, Herrmann, Radig and Lames (2016).

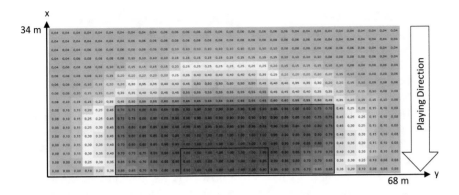

Fig 3.1: Quantification of Zone is carried out using the position of the IBA-player on a 2 m × 2 m grid that begins 34 m from the goal line.

Dangerousity is based on the four components *Zone* (ZO), *Control* (CO), *Pressure* (PR) and *Density* (DE), where the first two components increase and the last two components decrease its value. Zone represents the danger of a goal being scored from the position of the IBA-player, Control stands for the extent to which the player can implement his tactical intention on the basis of the ball dynamics, Pressure represents the opportunity of the defending team to prevent the IBA-player from completing an action with the ball and Density is the chance of being able to defend the ball after the action. The value range for all of the constructs is between 0 (low) and 1 (high).

These individual components give the Dangerousity for a moment t as the product of Zone and a linear combination of Control, Pressure and Density (Eq. 3.1). The model constant k_1 quantifies the extent to which these three figures reduce the value for Zone. It is selected in such a way that Zone is reduced by a maximum of a factor of 0.5. As a lack of control of the ball results in a reduction in the Dangerousity, Control is included as negated.

$$DA(t) = ZO(t) \cdot (1 - \frac{1 - CO(t) + PR(t) + DE(t)}{k_1}) \qquad \text{(Eq. 3.1)}$$

Quantification of Zone is carried out using the position of the IBA-player on a 2 m × 2 m grid that begins 34 m from the goal line (Fig. 3.1). Our evaluation of a position is based on several assumptions: First, as the distance from the goal decreases and centrality increases, the danger rises (Lucey et al., 2015; Pollard, Ensum & Taylor, 2004). Second, moving into the penalty area brings about a sudden increase in the danger because of the risk of a penalty kick (Tenga, Holme, Ronglan & Bahr, 2010). Third, there is a homogeneous area in front of goal in which the danger does not increase further. Fourth, an acute angle to the goal reduces the danger. Fifth, areas to the side of the penalty area are dangerous because of the possibility of a cross with little risk of offside.

Control is estimated by means of the average relative speed (v_{rel}) of ball and IBA-player in the last 0.5 s. High relative speeds occur, for example, when the player shoots on goal with just brief contact with the ball after a cross. Comparatively low relative speeds are found when the player has the ball at his feet for a longer period, when dribbling, for example, or positioning the ball for a shot on goal. We believe that at relative low relative speeds, there is an almost perfect CO near a value of 1. With increasing v_{rel}, it gets more and more difficult to control the ball. We model this by using a quadratic function, moderated by the model constant k_2 (Eq. 3.2). If v_{rel} is above $25 \, \mathrm{m\,s^{-1}}$, CO is equal to 0.

$$CO(v_{rel}) = 1 - k_2 \cdot v_{rel} \qquad \text{(Eq. 3.2)}$$

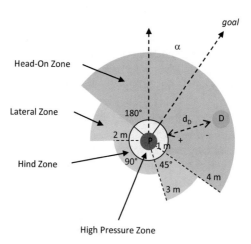

Fig 3.2: Geometry to determine Pressure. The Pressure Zone covers four sub-areas with different radii, which result from the angle (α) between IBA-player and the center of the goal. Pressure depends on the sub-area and the distance (d_D).

In determining Pressure, we assume that a *defender* (D) exerts pressure when his distance (d_D) from the IBA-player (P) is below a threshold value r_{ZO}. The *Pressure Zone* (PZ) covers four sub-areas with different radii (r_{ZO}), which result from the angle (α) between IBA-player and the center of the goal (Fig. 3.2). This is based on the assumption that a defender who is between the IBA-player with the ball and the goal (Head on Zone) is more likely to be able to defend a scoring opportunity than a defender who is to the side (Lateral Zone) or behind (Hind Zone). Within the zones, there is a linear increase in pressure as the distance falls. If the defender is very close to the IBA-player (High Pressure Zone), the pressure is constantly high. An individual defender (D_i) creates Pressure (PR_{Di}) in accordance with Eq. 3.3. Our model bases on the idea, that every additional defender increases Pressure, although the increase gets less with every additional defender. We model this using a logarithmical function, moderated by the model constant k_3 (Eq. 3.4).

$$PR_{D_i}(d_{D_i}, \alpha) = 1 - \frac{d_{D_i}}{r_{ZO}(\alpha)} \qquad \text{(Eq. 3.3)}$$

$$PR(x) = 1 - e^{-k_3 x}, \text{where } x = \sum_{\forall \ D_i \ inside \ PZ} PR_{D_i} \qquad \text{(Eq. 3.4)}$$

Density is described by means of two components: *Shot Density* (SD) represents the probability of a team blocking a shot, while *Pass Density* (PD) indicates the likelihood of intercepting an offensive pass or cross. Depending on the *Centrality* (C) of the IBA-player, the two components are weighted differently (Eq. 3.5). At an acute angle to the goal, Pass Density is weighted higher, in a central position the Shot Density is greater.

$$DE(c) = C \cdot SD + (1 - C) \cdot PD \qquad \text{(Eq. 3.5)}$$

A defender increases Shot Density if he is in the *Blocking Zone* (BZ) formed between the position of the IBA-player and the goal (Fig. 3.3). The value is calculated from the distance between the IBA-player (P) and the goal (d_{goal}) and between the IBA-player and the defender (d_{Di}). The smaller d_{Di} is, the higher the density created by that player because a larger area of the goal is potentially covered. For a defender (D_i), the SD_{Di} created by him is given by Eq. 3.6. Every additional defender within the BZ increases the density, although the increase is also attenuated logarithmically similar to Eq. 3.4, but with using a different model constant k_3.

$$SD_D(d_D, d_{goal}) = 1 - (d_D)\frac{d_D}{d_{goal}} \qquad \text{(Eq. 3.6)}$$

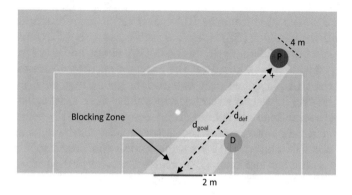

Fig 3.3: Geometry to determine Shot Density. A defender increases Shot Density if he is in the Blocking Zone formed between the position of the IBA-player and the goal. The figure is calculated from the distance between the IBA-player (P) and the goal (d_{goal}) and between the IBA-player and the defender (d_D).

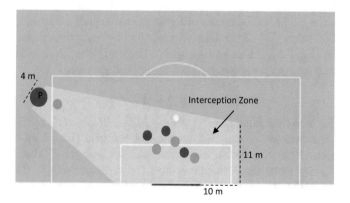

Fig 3.4: Geometry to determine Pass Density. Pass Density depends on the difference between the number of defenders and attackers (Majority) within the Interception Zone (IZ).

Pass Density depends on the difference between the number of defenders and attackers within the *Interception Zone* (IZ). We call this difference *Majority* (M). For example, if there are 4 defenders and 3 attackers in the IZ, M is equal to 1 (Fig. 3.4). As the Majority of defenders increases, Pass Density approaches a value of 1, with a Majority of attackers it moves towards 0. This understanding is operationalized by using an arcus tangent function (Eq. 3.7). The model constant k_5 describes the sensitivity of the model.

$$PD(M) = 0.5 + \frac{\tan^{-1} \cdot (k_5 M)}{\pi} \qquad \text{(Eq. 3.7)}$$

3.2.2 Derived Metrics

Dangerousity is present in the measuring frequency of the tracking system and can therefore be used to describe the value of IBAs. We call this quantity *Action Value* (AV) (Eq. 3.8). For calculating this, we use the difference between Dangerousity at the moment when a player has IBA (Start IBA) and the moment when the next player has IBA. If this is a player of the opposing team or DA decreased during the IBA, Action Value is negative.

$$AV(IBA_n) = DA(Start\ IBA_{n+1} - DA(Start\ IBA_n) \qquad \text{(Eq. 3.8)}$$

In order to assess the success of the attack, which we call *Performance* (PE), of a team over a longer period, the match is divided into intervals of 5 s in length and the maximum value for DA is determined for both teams over an interval i (DA_i). Performance is then given by the sum of this value for all intervals over the period *ts*, as *Match Performance* (MP), for example (Eq. 3.9). This discretization ensures that there are the same number of summands for both teams over a given period. Furthermore, we use *Current Performance* (CP) in order to describe the course of play over a time interval. We determine this for a moment t using past values for DA_i in the intervals of the last 5 minutes.

$$PMP(ts) = \sum_{\forall i:\ Interval_i\ \in\ ts} \cdot DA_i \qquad \text{(Eq. 3.9)}$$

While the previous metrics are based on an evaluation of the attacking play of a team, *Dominance* (DO) describes the difference in performance between the two teams (T1, T2). This can be calculated both over a time interval as *Match Dominance* (MD), for example, and for a moment as *Current*

Dominance (CD). In both cases, this is provided by the difference in the Performance of the two teams (Eq. 3.10).

$$DO(T1) = PE(T1) - PE(T2) \qquad \text{(Eq. 3.10)}$$

3.2.3 Calibration

The calibration processes intended to optimizes the model constants k_1, k_2, k_3, k_4, k_5 manually. In collaboration with soccer experts, a large number of different match situations were analyzed in detail, such as positional attacks, counter-attacks, 1 vs. 1 situations, crosses and dribbling with the ball by individual players. With the aid of a self-developed software package, the individual match scenarios were compared with the quantification of the components in the dangerousity model. In this process, it was possible to simulate the effect of changing model constants and to optimize them gradually. A total of over 100 situations were examined for apparent validity (see example in subsection 3.4.1).

3.2.4 Limitations

In balancing complexity, the accuracy of the positional data and the benefit for performance diagnostics, the procedure presented here does not take account of all aspects of dangerousity. These include the movement dynamics of the players and the ball, the direction in which the players are looking, their position in relation to the ball, the extent to which teammates are available (Gudmundsson & Wolle, 2014; Kang et al., 2006) and different individual skills. Also all geometrical parameters base on our qualitative evaluation of game situations, our interpretation and – at the end – on our philosophy of the game. Further studies should empirically validate some assumptions, e.g. the model for Zone. The treatment of special cases such as standard situations, off sides and retrospective sanctions for fouls is out of the papers scope.

3.3 Evaluation

3.3.1 Reliability of measurement

This paper is based on 64 games in the German Bundesliga (2014/15 season). The positional data was collected by a CIP (TRACAB corp.) via an optical tracking system and then reviewed manually. Since each player agreed to this procedure on signing his contract of employment as a professional soccer player, special approval for this study from an ethics committee was not required. Nevertheless, all procedures performed in the study were in strict accordance with the Declaration of Helsinki as well as with the ethical standards of the Chair of Training Science and Sports Informatics of the Technical University of Munich.

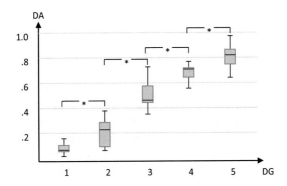

Fig 3.5: Boxplot of Danger. Match scenarios (n = 100) were grouped into Danger Groups (DG) by experts using a scale of 1 (little danger) to 5 (very dangerous). Scenarios that were classified as dangerous by the observers were also classified as dangerous on average by the algorithm. All of the post hoc tests between neighboring groups also showed significant differences ($\alpha = .01$).

In assessing the quality of the quantification, the automatic calculation of dangerousity was compared with the evaluation by semi-professional soccer coaches in 100 match scenarios. The sample was selected in such a way that the value range of DA was covered evenly. Three experts evaluated the scenarios independently of one another on the basis of video recordings using a scale of 1 (little danger) to 5 (very dangerous). They had no knowledge of the underlying model, but were asked to evaluate the scenarios qualitatively in their entirety. For the statistical analysis, we grouped the situations into *Danger Groups* (DG) following their assessments by the majority principle and checked for differences using an ANOVA.

The results show that the mean value for DA differed significantly between the groups (F = 170.31, p < .01) (Fig. 3.5). All of the post-hoc tests between neighbouring groups also showed significant differences $\alpha = .01$). This means that scenarios that were classified as dangerous by the observers were also classified as dangerous on average by the algorithm. In some cases, the classifications of the observers differ from one another, but also between an observer and the algorithm. There is a fair correspondence between the observers (κ = .32; Fleiss, 1971). This is also to be expected, as the quantification of danger also includes subjective components.

3.3.2 Performance relevance

The crucial criterion for the quality of a performance indicator is that it depicts an issue that describes an important component in the performance of a sport. This is usually checked in the performance analysis in two different ways. One possibility is to determine its capacity to forecast the outcome of a match (Castellano et al., 2012; Lawlor, Low, Taylor & Williams, 2003). Since, however, relatively few goals are scored in soccer, as discussed in the introduction, the number of goals is usually only moderately related to performance indicators (Tab. 3.1). A more promising approach is therefore to assess the contribution of an indicator to clarification of performance – i.e. the ongoing playing strength of a team – and to compare this with other indicators (Lago-Ballesteros & Lago-Peñas, 2010; Oberstone, 2009).

For the matches in the sample, the variables *Goal* (G), *Shot at Goal* (SG), *Passing Accuracy* (PA), *Tackling Rate* (TR), *Ball Possession* (BP) and *Match Dominance* (MD) were collected. G, SG and PA represent the difference in the variable value from that of the opponent; the remaining variables already represent relative values between the teams. As an external criterion for the difference in performance between the teams, the *Win Probability* (WP) of a team was used, based on the odds from 13 bookmakers (www.Football-Data.co.uk). In the statistical analysis, the correlation coefficients for all the pairs of variables were calculated.

The highest correlation between WP and the performance indicators exists with MD (r = .82), followed by BP, SG, PA, G and TR (Tab. 3.1). Of the indicators studied, dominance is therefore the one with the highest correlation with the performance of a team. It follows from this that the probability of a stronger team in a match achieving a higher dominance is greater than the probability of it scoring more goals, for example. This is easy to explain in terms of content, as a team that is weaker in a match is more likely to score a goal from an individual situation than it is to generate more dangerous situations over the entire course of a game. In this context, the sequence given above can be taken as a way of sorting the indicators according to their relevance for match performance. The validity of domin-

Tab. 3.1: Correlations between performance indicators and between performance indicators and skill indicator (win probability (WP) based on
betting odds). The greatest correlation between WP and the performance indicators exists with Match Dominance (MD).

	Skill indicator	Performance indicator					
	WP	G	SG	PA	TR	BP	MD
G	.55	x	.44	.33	.35	.34	.41
SG	.58	.44	x	.64	.20	.61	.83
PA	.56	.33	.64	x	.26	.93	.78
TR	.24	.35	.20	.26	x	.19	.14
BP	.71	.34	.61	.93	.19	x	.76
MD	.82	.41	.83	.78	.14	.76	x

ance and therefore, in turn, of performance can thus be demonstrated both
rationally and empirically-quantitatively.
Other evidence of validity emerges from the correlation of performance indicators with one another. Here the results seem entirely plausible: Shots
on goal are completed mainly in situations with a high dangerousity, so
there is a strong correlation between MD and SG ($r = .83$). As Dangerousity presupposes possession of the ball, MD and BP also show a strong
correlation ($r = .76$). Possession of the ball and passing accuracy are almost
entirely identical (r = .92). This is easy to understand because a poor pass
leads directly to the loss of the ball.

3.4 Game Analysis

3.4.1 Individual Action Analysis

A possible application at the micro level is the analysis of small sections of
game situations. As dangerousity is calculated for every frame, the value
changes continuously during an IBA interval. Key situations such as out-
playing an opponent in an important duel or a successful pass through
the defensive line cause a big jump in DA, while periods without gaining
ground lead to an even signal sequence. This can be illustrated by an attack
by Bayern Munich (FCB) against TSG 1899 Hoffenheim from the Saison
2014/2015 (Fig. 3.6).

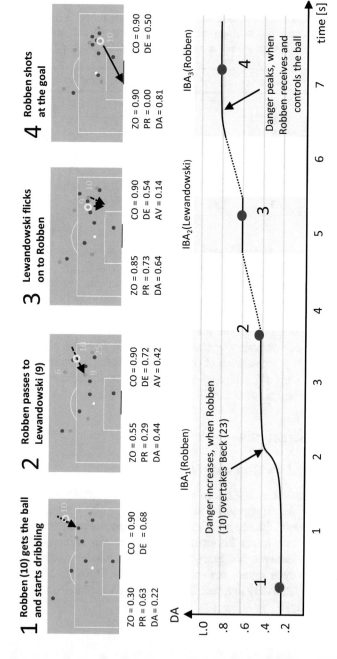

Fig 3.6: Course of Danger in a match scenario. Spatial configuration and value of model components are shown in four key moments.

The attack comprises three IBA intervals by the players Robben (No. 10), Lewandowski (No. 9) and Robben again. At the moment when Robben takes possession of the ball, there is moderate dangerousity ($DA = .22$). The player begins to dribble against defender Beck (No. 23) and he succeeds in beating his opponent. This results in a reduction in PR, as Beck falls back from the High Pressure Zone into the Hind Zone. At the same time, ZO increases because of the shorter distance to the goal. At the end of IBA, there are three realistic play options for Robben, apart from continuing to dribble. The pass options to Müller (No. 25) and Lewandowski would be evaluated almost equally by the Action Value ($AV_{IBA\_1} = .44$ and $.42$), assuming that the spatial configuration does not change significantly, while the back pass to Thiago (No. 6) would result in a negative evaluation ($AV_{IBA\_1} = .19$). Robben decides to pass to Lewandowski and gets the ball straight back from the player ($AV_{IBA\_2} = .14$). After contact with the ball to take possession of it, DA increase to its maximum value in this scenario ($DA = .81$). There is then a stand-alone shot on goal by Robben, but from a relatively acute angle in front of the goalkeeper.

The diagnostic benefit in terms of performance of this analysis lies less in the evaluation of the playing behaviour in individual scenarios. This would require a multitude of other factors to be taken into consideration, such as movement dynamics, passing risk and individual skills, which could only be derived from a qualitative analysis of the film material. By contrast, it would be possible to evaluate the contribution of a player to the attacking play of his team over a longer period. It is possible that players will thus be identified who, although they have a lot of contact with the ball, contribute only little to increasing dangerousity in attacking phases.

The key application for the game analysis lies in filtering video material on the basis of dangerousity. Sudden increases can be understood as disruptions or perturbations in the balance between defence and attack in line with the theory of dynamic systems (Hughes, Dawkins, David & Mills, 1998; James et al., 2012). The specific selection of these scenarios, possibly in combination with other attributes such as the side of the pitch or the involvement of certain players, can simplify game analyses significantly.

3.4.2 Single match analysis

Traditional match statistics provide inadequate information to assess the course of a match correctly (Lucey et al., 2015; Mackenzie & Cushion, 2013). As already shown in the introduction using the example of BRA - GER, shots on goal are not very suitable as a criterion for assessing performance or success in specific cases. The same applies to possession of the ball: teams that are in the lead often change their tactics and then have less possession of the ball than if they are behind (Jones et al., 2004; Lago-Peñas & Dellal, 2010). Tackle and pass rates show only small links with the performance of teams (Castellano et al., 2012, Lawlor et al., 2003; Tab. 3.1). We therefore believe that performance or dominance allow a significantly better assessment of whether a team has been 'lucky' and won through an individual action or has been able to set up many dangerous situations and has 'earned' the win.

Match 1 provides an example of a merited victory by the home team (Fig. 3.7). Here, Dortmund (BVB) creates significantly more dangerous situations (PE 466:138), despite even possession of the ball. The opposite course of events can be assumed in match 2: Schalke (S04) dominates the match with a PE of 469:198, but suffers a defeat by 2:1. In match 3, although Wolfsburg

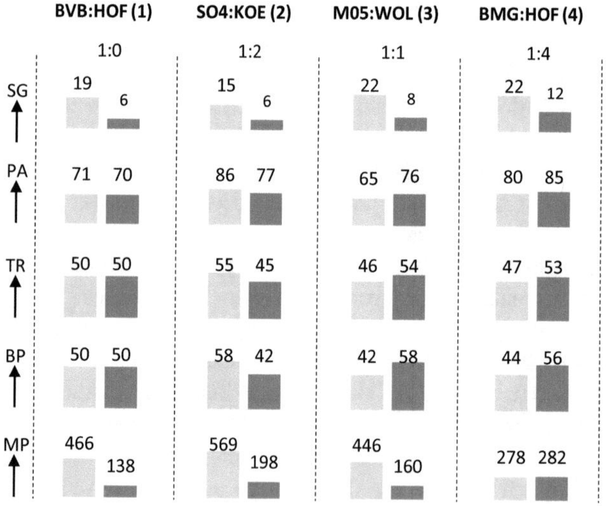

Fig 3.7: Performance indicators of 4 matches. Goals (G), Shots at Goal (SG), Passing Accuracy (PA), Tackling Rate (TR) and Ball Possession (BP) provide inadequate information to assess the course of a match correctly. Match Performance (MP) allow a significantly better assessment of whether a team has been 'lucky' and won through an individual action or has been able to set up many dangerous situations and has 'earned' the win.

(WOL) has more possession, it creates significantly lower PE from it than Mainz (M05). This constellation suggests a large number of unsuccessful positional attacks by a team that also has problems preventing their opponents from counter-attacking. In match 4, Gladbach (BMG) has more shots on goal than its opponent but without achieving an advantage in PE to the same degree. The shots on goal may have come from situations involving comparatively little danger. As far as pass and tackle rates are concerned, we do not believe that any plausible relationships are evident.

Using the metrics Current Performance and Current Dominance, success can be assessed not only for a complete match, but also for periods of time. In this way, the effects of tactical interventions (substitutions, system changes) or central events in a match can be investigated, for example. Fig. 3.8 shows the course the match between Hannover 96 (H96) and Borussia Dortmund (BVB) on the 26th match day. With the metrics, we manually identified eight key phases (P). Until the first goal is scored (0:1, 19th min), the play can be described as very even (0 – 20th min). Then a phase of dominance (20th - 36th min) began for H96, during which they took the score to 1:1 (25th min). Between the 36th and 42th minutes, BVB managed only the occasional attack. In the last 4 min of the first half, H96 clearly dominates play but without scoring another goal.

At the start of the second half (45th - 57th min), play was very even up to the sending-off for H96 (55th min), with a slight advantage to H96. The dominant phase for BVB (57th - 76th min) began between their second goal (1:2, 57th min) and their third (1:3, 61st min). This may have been the result of their superiority in numbers and/or psychological elements. This phase ended around 15 minutes before the end of the match. After conceding their second goal (2:3, 82nd min), BVB recorded no further successful attacks and obviously attempted to play out time. H96 was able to create a further series of dangerous plays, but without levelling the score. If dominance is considered over the two halves, it is clear that the home team was dominant after the first half (DO = 90), while the away team had the advantage in the second half (DO = −41).

3.4.3 Team efficiency ratings

At the macro level, performance and dominance are appropriate for characterizing the performance of the team as a whole. It is important firstly to consider the attacking performance and the defensive performance (as the inverse of the opponent's attacking performance) over a large number of matches and to put these in the context of the other teams. This is a good

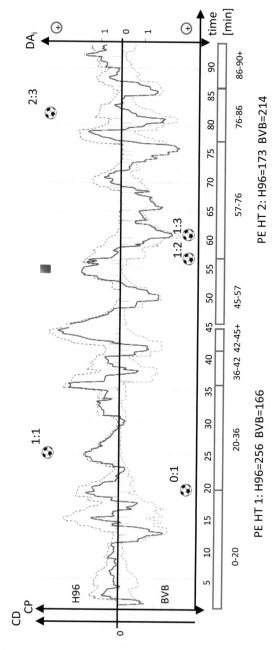

Fig 3.8: Performance variables in the course of the match Hannover (H96) vs. Dortmund (BVB). Danger for an interval (DA$_i$) is visualized by bars, Current Performance (CP) by dashed lines and Current Dominance (CD) by a solid line. DA$_i$ and CP were inverted for the away team. CD is shown from the perspective of the home team.

starting point for describing team efficiency, which can be defined as the
relation between points achieved and the dominance. This allows to answer
the question of the extent to which the examples of lucky victories shown
in Fig. 3.7 are balanced out by unlucky defeats over the course of a season,
for example.

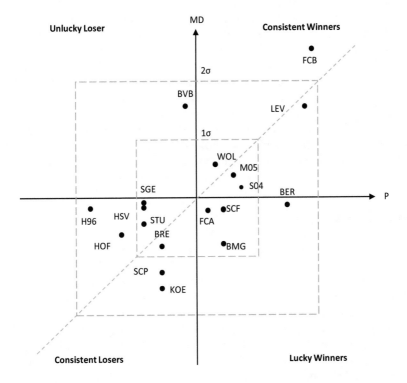

Fig 3.9: Ranking of teams based on 64 matches. The vertical position of a
team indicates its average Match Dominance (MD), the horizontal position describes the number of points (P) it has gained. The quadrants
classify team according to the factors successful vs. less successful and
consistent vs. lucky. A position above the main diagonal indicates an
unfavorable ratio of Dominance to success (Points), a position below
it a positive ratio.

Even though we have only 64 matches at our disposal for our study, the
principle of the procedure can be illustrated in Fig. 3.9. Here the vertical
position of a team indicates its average dominance, the horizontal position

describes the number of points it has gained. In the 1st quadrant there are teams that both dominated in matches and have had above-average success. Teams in the 2nd quadrant dominated but were less successful. Similarly, the subordinate teams in terms of play can be divided into successful (3rd quadrant) and unsuccessful (4th quadrant).

If one interprets the main diagonal as the number of points expected for the level of dominance, the horizontal distance of a team from this diagonal can be understood as the discrepancy between effort and success. A position above the main diagonal indicates an unfavorable ratio of dominance to success, a position below it a positive ratio. In other words, the teams below the diagonal were quite efficient, the teams above were not. In the matches examined, for example, the teams from Dortmund (BVB) and Hanover (H96) were significantly less successful than they 'deserved' to be based on their dominance. Berlin (BER), Gladbach (BMG) and Cologne (KOE), on the other hand, can be happy with their points return in view of their match performance. The success of Leverkusen (LEV) and Mainz (M05) roughly corresponds to their match performance.

3.5 Conclusion

The aim of the study was to develop, evaluate, and apply a procedure for determining dangerousity in soccer with real-time capability. The evaluation showed that the quantification of this construct using the spatial constellation of players and ball lies in the same range as human observers. Individual misinterpretations play only a subordinate role in the validity of diagnostic findings concerning performance, particularly with large data volumes. Like many other tactical elements in soccer, however, the construct does contain a certain lack of precision, with the result that a clear reference for the accuracy of the measurement in the sense of a ground truth cannot exist.

The performance and dominance metrics derived are more robust in the context of the effects of chance, and map the match performance of a team more reliably than the traditional performance indicators of possession of the ball, shots on goal, tackle, and pass rates. They can be used to evaluate individual plays, to describe efficiency, represent passages of play, or compare players and teams with one another. In particular, they can help to investigate questions relating to the influence of various playing systems or tactical group concepts on success. In addition, the metrics can be used as the basis for the development of media products, e.g. the fever curve in

Fig. 3.8 could be shown during television broadcastings, live event tickers, or second screen applications.

Acknowledgments   This chapter is taken from Link, D., Lang, S. & Seidenschwarz, P. (2016). Real time quantification of dangerousity in football using spatiotemporal tracking data. *PLOS ONE*, *11*(12), e0168768. doi:10.1371/journal.pone.0168768.

4 A Topography of Free Kicks

Abstract This study investigates the spatial relationship of performance variables for soccer free kicks. In order to suggest ways in which players might optimize their performance, we collected data from free kicks ($< 35\,\text{m}$ to goal line) of two German Bundesliga seasons (2013/14, 2014/15) (n = 1,624). In the analysis, we applied the ISO-map approach using color gradients to visualize the mean values of a variable on a 2D-map of the pitch. Additionally, variograms were used to describe the degree of spatial dependence of the free kick variables. Results show that DENSITY, TYPE OF PLAY, PLAYERS IN WALL, DISTANCE TO WALL and RULE VIOLATION were strongly spatially dependent. Centrality and proximity to the goal increased the variables PLAYERS IN WALL, RULE VIOLATIONS and INTERRUPTION TIME, and the ratio of goals scored increased from 5.9 % (central far) to 10.9 % (central near). In 70.9 % of the shots, players preferred a switched laterality, which did not result in a higher success rate. Furthermore, there was no statistical advantage for the defensive team when DISTANCE TO WALL was below 9.15 m or when there was a RULE VIOLATION. Crosses had a success rate (i.e. first controlled ball contact after the cross) of 20.8 %. Played with natural laterality, they were 5 % more successful than with switched laterality. Crosses from the right side outside the penalty box were 10 % more successful than from the left side. Therefore, it might be worthwhile practicing the defense of balls coming from this side.

4.1 Introduction

Free kicks represent an important component of performance in soccer. Previous research has found an average number of 30-40 per game (Casal, Maneiro, Ardá, Losada & Rial, 2014; Ensum, Williams & Grant, 2000; Hernández Moreno et al., 2011; Olsen, Larsen, Reilly, Bangsbo & Hughes, 1995; Siegle & Lames, 2012) and that they are the most effective set play from which goals are scored (Carling, Williams & Reilly, 2005). Analyses of World Cups from 1982 to 2014 showed that between 25 % and 35 % of goals result from free kicks or corner kicks (Acar et al., 2009; Grant et al., 1999; Jinshan, Xiaoke, Yamanaka & Matsumoto, 1993; Njororai, 2013; Yiannakos & Armatas, 2006; Yiannis, 2014). Furthermore, successful teams needed fewer set plays to score goals than less successful ones (Bell-Walker, McRobert, Ford & Williams, 2006).

Existing studies have sought to identify the factors that increase the probability of successful shots or crosses resulting from set plays. Regarding shots on goal, the chances of success clearly increased for high-velocity shots (Kerwin & Bray, 2006), as well as for positions close to the goalposts (Acar et al., 2009; López-Botella & Palao, 2007; Morya, Bigatao, Lees & Ranvaud, 2005). The results are not as conclusive with regard to free kicks played as crosses. Carling et al. (2005) reported that crosses directed towards the center of the penalty area or in the zones in front of the goalposts increase the chances of scoring, whereas Casal et al. (2014) could not find any correlation between position and scoring. A study by Taylor, James and Mellalieu (2005) suggests that corner kicks are more successful when they are played near the six-yard line. Furthermore, out-swinging corner kicks led to more shots and in-swinging corner kicks to more goals.

To date, there have been relatively few findings on the effects of the position from which a free kick is executed and the characteristic variables related to this position. Only two studies exist that have classified free kicks into distinct groups depending on their location and then described differences between them. Firstly, Alcock (2010) reported that, during the 2007 Women's World Cup, there was an increased scoring rate for direct free kicks if these were taken within a 7 m radius of the penalty spot. However, the small sample of only 65 free kicks does not seem sufficient to derive a general statement in this respect. The second study considered 783 free kicks played as crosses during the 2010 World Cup and the 2012 European Cup (Casal et al., 2014). In this study, the position of the free kick could not be found to influence either the scoring rate or the shot rate. Additionally, laterality was not found to influence success either. Due to the discretization of free kick positions, neither of these studies provides detailed conclusions on the continuous profile of the variable values.

However, continuous profiling approaches exist in other scientific disciplines, such as geostatistics. For instance, *variography* (Isaaks & Srivastava, 1989) examines the spatial dependency of parameters and is mostly used in combination with prognosis models. Applications can be found in mining (Samal, Sengupta & Fifarek, 2011), hydrology (Lepioufle, Leblois & Creutin, 2012) or ecology (Reis, da Silva, Sousa, Patinha & Fonseca, 2007). In sport, a similar approach has been used for the analysis of golf shots

(Stöckl, Lamb & Lames, 2012), and to describe patterns of play in hockey (Stöckl & Morgan, 2013).

In the light of the above, this study aims to provide a continuous spatial description of free kick parameters and to identify their influence on success. Some of these parameters, such as execution type, interruption period and laterality have already been examined, though with a different focus and/or with smaller sample sizes. Other parameters used in this study such as the number of players in the wall formation, distance to the wall formation and rule violations of players in the wall have not previously been taken into account. The current study thus allows new insights into the performance structure of soccer and suggests ways in which coaches and tacticians could improve and optimize the performance of their players.

4.2 Methods

4.2.1 Sample

In line with the objectives of our study, we applied a non-participative observational approach. The sample comprises all free kicks from all 612 matches of the German Bundesliga seasons 2013/14 and 2014/15. Free kicks were registered by a professional competition information company (Opta corp., London) and provided by the German Professional Soccer League (DFL).

4.2.2 Performance Variables

For all free kicks, POSITION was collated by the competition information provider. In order to describe the frequency rate of free kicks, we divided the pitch into $1\,\mathrm{m}^2$ sectors. DENSITY represents the number of free kicks counted in each sector. Additionally, for each free kick with a distance of less than 35 m to the goal line, the following performance parameters were collected (Tab. 4.1): INTERRUPTION TIME, DISTANCE TO WALL, PLAYERS IN WALL, RULE VIOLATION, LATERALITY, TYPE OF PLAY, OUTCOME OF SHOT and OUTCOME OF CROSS.

DISTANCE TO WALL was measured in the last four rounds of the 2014/15 season (encompassing 36 matches) using video footage containing a view of the entire pitch without camera panning and homography. All other variables were observed by two trained human experts based on video recordings. Due to video availability and cutting, some free kicks could not be

Tab. 4.1: Performance variables, categories and operationalization.

Variable	Value
POSITION	2D-position of free kick (m). Coordinates are normalised to playing direction of executing team.
ZONE	Free kick location grouped according to Fig. 4.2
DENSITY	Number of free kicks in a $1\,\mathrm{m}^2$ sector.
INTERRUPT-ION TIME	Timespan (s) between the foul that causes the free kick and the moment of ball contact. Only collected for free kicks where vanishing spray was not used.
DISTANCE TO WALL	Shortest distance (m) between ball and wall in the moment of ball contact.
PLAYERS IN WALL	Number of players participating in the wall.
RULE VIOLATION	*true*: At least one player inside the wall tries to reduce distance to the ball for more than $30\,\mathrm{cm}$ after the referees allows the execution of the free kick.
	false: Otherwise
LATERALITY	*natural*: For crosses: ball was played from the left side with left foot or from the right side with the right foot. For shot at goal: Ball was played from central left with the left foot or from central right with the right foot.
	switched: Otherwise
TYPE OF PLAY	*shot at goal*: Player intends to score a goal directly with the ball contact.
	cross: Player plays the ball from outside into the penalty box. Ball trajectory has a length of at least $10\,\mathrm{m}$.
	pass: Otherwise

OUTCOME OF SHOT	*goal*: Goal was scored.	
	border: Ball touched the border of the goal.	*on target*
	saved: Ball was caught, fisted or defended by the goalkeeper.	
	off target: Ball passes the goal line outside the goal.	*missed*
	block outside: Ball was blocked by player outside the wall.	
	block: Ball was blocked by player inside the wall.	*wall*

OUTCOME OF CROSS	*shot at goal*: Player intends to score a goal with the first ball contact after the cross.	
	flick-on: Controlled flick-on in goal direction or to team-mate.	*successful*
	no control: Ball was touched by the attacking team without control.	
	opponent: Ball was touched by the opponent team.	*not successful*
	block: Ball was blocked by player inside the wall.	

considered in our study. The quality of the data was evaluated by an inter-rater reliability test based on the two observers. Cohen's kappa statistics showed substantial to almost perfect agreement ($\kappa = .69$ up to .90).

4.2.3 Calculation of performance topographies

An approach introduced by Stöckl and Lames (2012) was used to visualize the patterns of the different performance variables. This method provides a continuous topography, a so-called *ISO-map*, of the investigated perform-ance characteristic. The topography values are an approximate average of the measurements and are generated in four steps:

1. A grid is put on the field of play. The grid size was 0.5 m in this study.

2. At each grid node, a representative value is calculated using exponential smoothing. To ensure that the representative value is based solely on close measurements, a small smoothing factor had to be chosen. In this study, the smoothing value was 0.2 for all performance variables, meaning that about the 10 closest measurements to a grid node are considered.

3. The values between the grid nodes are interpolated using a cubic smooth-ing spline. A relatively strong smoothing value of 0.1 was used for all performance variables.

4. Two-dimensional representations of the performance variables were de-termined. They illustrate discrete levels of the values of the respective performance variable, which are color coded.

Generally, this approach requires triplets (x,y,z), where (x,y) denotes the location of a measurement, and z denotes the measured value of a perform-ance. All performance variables provided measurements consisting of the location and a performance characteristic, either measured as part of a ra-tio scale (INTERRUPTION TIME, DISTANCE TO WALL, PLAYERS IN WALL, DENSITY) or as binary value (RULE VIOLATION, TYPE OF PLAY, OUTCOME OF SHOT, OUTCOME OF CROSS). For the ratio scaled parameters, the ISO-maps illustrate the raw values. For the binary parameters, one parameter was set 1, such as goal, and the other 0, such as no goal. The resulting topographies, then, visualize the percentage of cases fulfilling the parameter option which was set 1.

The ISO-map calculations were performed using MATLAB R2014a (The MathWorks Inc., USA).

4.2.4 Statistical Analysis

One of the aims of this study was to investigate the spatial dependence of the different performance variables. For this purpose, the well-known geostatistical approach variography (Isaaks & Srivastava, 1989) was used. This studies the spatial dependence of a variable using variograms. A *variogram*, also called *semivariogram*, is a graph representing a function which describes the degree of spatial dependence between measurements at sample locations. Variography is conducted in several steps. First, an empirical semivariogram is calculated based on the sample data z_i as

$$\gamma(h) = \frac{1}{2|N(h)|} \sum_{i,j \,\in\, N(h)} (z_i - z_j)^2 \qquad \text{(Eq. 4.1)}$$

where h is a small range of distances, and $N(h)$ denotes the set of observations which pairwise fulfil $|z_i - z_j| = h$ (Isaaks & Srivastava, 1989). Next, a theoretical variogram is fitted to the empirical semivariogram using different base functions. In particular, Gaussian functions (Gau) and Matern M. Stein's parameterization functions (Ste) ($\kappa = 10$) were used in this study.

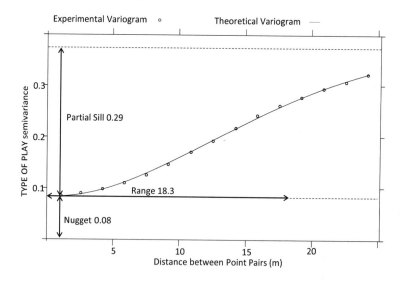

Fig 4.1: Experimental and theoretical variogram of TYPE OF PLAY.

The theoretical variograms provide the parameters nugget, sill and range. The *nugget* (c_0) describes the variability in the data at distance h = 0 and can be interpreted as the amount of variability which cannot be explained by the location in the measured area and/or are measurement errors. A nugget value of zero indicates that the location completely explains the variability. The *sill* (C) describes the maximal semi-variance, which can be interpreted as there being no (spatial) correlation. The *range* is the distance within which the values are correlated. Furthermore, the *nugget-to-sill ratio* (NSR = $\frac{c_0}{C}$) can be used to quantify the spatial dependence of a studied property. According to Cambardella et al. (1994), a NSR smaller than .25 means a strong spatial dependence, .25 − .75 suggests moderate spatial dependence and a value > .75 denotes weak spatial dependence.

Fig. 4.1, for example, shows the variogram of one of the parameters investigated, TYPE OF PLAY. The circles illustrate the empirical variogram, and the line fitted to it is the theoretical variogram based on a Gaussian function. There is a nugget of .08 and a partial sill of .29. Those values add up to the sill, which is .37. Thus, the nugget-to-sill ratio is .22. The range of 18.3 m indicates that the execution type of free kicks was correlated with one of the other free kicks, which are 18.3 m away at maximum. In total,

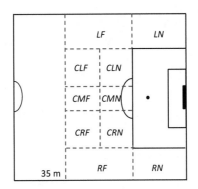

Fig 4.2: Discrete ZONEs of free kicks based on their POSITION: Left near (LN), left far (LF), left = {LN, LF}, right near (RN), right far (RF), right = {RN, RF}, far = {LF, RF}, near = {LN, RN}, central left near (CLN), central mid near (CML), central right near (CRN), central near = {CLN, CMN, CRN}, central left far (CLF), central mid far (CMF), central right far (CRF), central far = {CLF; CMF, CRF}, central left ={CLN, CLF}, central right ={CRN, CRF}.

whether a free kick was executed as cross or as shot strongly depended on the location on the field of play.

In addition to geostatistical analyses, we used inferential statistics based on spatial groups. We defined discrete ZONEs (Fig. 4.2) and determined differences in the outcome of shots and crosses between these zones using χ^2-tests. In order to increase the statistical power, we combined outcome categories. For OUTCOME OF SHOT we used two sets of categories: Firstly, *on target*, *missed* and *wall* and, secondly, *goal* and *no goal* according to Tab. 4.1. For OUTCOME OF CROSS, we tested the categories *successful* and *non-successful*. Before using the test procedures, the assumptions of normality were verified. The alpha level was set to .05. The statistical analyses were performed using R; in particular, the package *gstat* (Pebesma, 2004).

4.3 Results

During both seasons there were 34.9 ± 7.6 (n = 21,414) free kicks on average per game. At a distance of less than 35 m from the goal line, there were 5.8 ± 2.48 (n = 3,579) free kicks. Tab. 4.2 shows the results of the variography. We examine the details of those results in the following when talking about the different parameters.

DENSITY of the free kicks strongly depended on the location (NSR = .18): There was a cluster in the penalty area of the defending team and also four lengthwise strips were observable in the midfield, in which the DENSITY was about $30 - 50$ % higher compared to the surrounding area (Fig. 4.3).

Tab. 4.2: Theoretical variograms determined for performance variables.

Indicator	Model	Nugget	Sill	Range[m]	NSR
DENSITY	Ste	3.0	15.7	173.8	.18
INTERRUPTION TIME	Gau	170.4	370.1	29.6	.46
DISTANCE TO WALL	Gau	0.7	3.1	79.0	.23
PLAYERS IN WALL	Gau	0.6	54.2	79.3	.01
RULE VIOLATION	Ste	0.1	0.5	90.1	.22
LATERALITY	Sph	0.2	0.3	227.2	.88
TYPE OF PLAY	Gau	0.1	0.4	18.3	.22
OUTCOME OF SHOT	Ste	0.1	0.1	1.3	.88
OUTCOME OF CROSS	Ste	0.2	0.2	5.7	.98

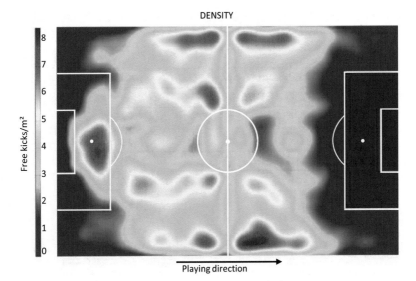

Fig 4.3: Spatial distribution of free kicks (DENSITY) (n = 21,414). Striking
are four horizontal lines with increased number of free (Bell-Walker,
McRobert, Ford & Williams, 2006) kicks count. The penalty box (offensive) was excluded by definition.

The strips along the sidelines in the first part of the attacking half were
especially pronounced. In the attacking last third of the pitch, the closer
players got to the penalty area, the fewer free kicks were registered.
In the sample examined (n = 2,681), 22.2 % of the free kicks were played
as shots (Tab. 4.3). For the remaining free kicks, the players brought the
ball back into play via a pass or a cross (both were equally as likely). The
influence of the location on the TYPE OF PLAY could be characterized as
strong (NSR = .22). Shots in particular took place from central positions
near the penalty area, while in the periphery of the field of play less than
about 10 % of the free kicks were played as shots (Fig. 4.4a). With regard
to LATERALITY, it was determined that 70.9 % of the shots at goal were
taken with switched laterality; for crosses this was only the case 56.1 % of
the time. If both execution types were considered together, then 71.0 % of
free kicks on the left side were taken with switched laterality, whereas on the
right side only 51.1 % were (Fig. 4.4b). This effect was weakly dependent
on POSITION (NSR = .88).

Tab. 4.3: Quantitative distribution (QD) of categories of TYPE OF PLAY, OUTCOME SHOT and OUTCOME CROSS (in %) (n = 2,681 free kicks within 35 m to the goal line).

Indicator	QD
TYPE OF PLAY	
shot at goal	22.2
cross	38.3
pass	39.5
OUTCOME SHOT	
goal	7.6
border	2.8
saved	27.7
off target	31.7
block	24.5
block outside	5.6
OUTCOME CROSS	
shot at goal	15.5
flick-on	5.1
no control	7.8
opponent	70.5
block	1.0

The ratio of goals scored from free kicks (OUTCOME OF SHOT) was 7.6 %, where 4.1 % of the 1,810 goals resulted from free kicks played as a shot (Tab. 4.4). An increase in the probability of a goal could be observed at around roughly 12 m in front of the penalty zone in the centre (Fig. 4.4c). Although the spatial correlation was weak (NSR = .88), a 5.0 % significant increase in probability of scoring could be observed for central and near free kicks. No significant effects of ZONE on the ratio of shots on goal, shots into the wall formation and shots that miss the goal were found. There were no differences between the probability of success from left and right, central free kicks played from near the penalty zone and laterality of free kicks.

The success rate of free kicks played as crosses (OUTCOME OF CROSS) was 20.6 % (Tab. 4.5). The ISO-maps (Fig. 4.4d) as well as the variography show that, as for shots, there was only a very weak correlation with the position of execution (NSR = .98). However, using group statistics, it was possible to show that there was a significant increase in success, by 10 %, for free kicks from the front right. This effect could not be observed for free kicks that were taken from further away. Free kicks that were played as

Tab. 4.4: Quantitative distribution of category OUTCOME SHOT grouped by spatial ZONE and LATERALITY (in %). Only free kicks in the listed zones were considered (n = 585). * indicates significance at level .05.

	n	on target	missed	wall	χ^2	p <	goal	no goal	χ^2	p
ZONE										
central near	356	35.1	38.8	26.1	1.13	.57	10.9	89.1	5.00	.03*
central far	229	33.7	36.2	30.1			5.9	94.1		
central left	234	30.3	38.9	30.8	1.38	.50	9.4	90.6	0.16	.92
central mid	112	34.8	34.8	30.4			8.9	91.1		
central right	239	37.7	36.4	25.9			8.4	91.6		
LATERALITY										
natural	138	39.1	31.9	29.0	3.35	.19	5.8	94.2	2.29	.13
switched	335	31.9	40.0	28.1			10.1	89.9		
natural left	43	34.9	30.2	34.9	1.73	.42	7.0	93.0	0.37	.54
switched left	190	28.9	41.1	30.0			10.0	90.0		
natural right	95	41.1	32.6	26.3	1.24	.54	5.3	94.7	1.95	.16
switched right	145	35.2	39.3	25.5			10.3	89.7		

Tab. 4.5: Quantitative distribution of category OUTCOME CROSS grouped by spatial ZONE and LATERALITY (in %). Only free kicks in the listed zones were considered (n = 1,028). * indicates significance at level .05.

		OUTCOME OF CROSS			
	n	Successful	Non Successful	χ^2	p
ZONE					
near	339	20.9	79.1	0.03	.86
far	689	20.5	79.5		
left	494	18.2	81.8	3.36	.07
right	534	22.8	77.2		
near left	155	15.5	84.5	5.22	.02*
near right	184	25.5	74.5		
far left	339	19.5	80.5	0.41	.52
far right	350	21.4	78.6		
LATERALITY					
natural	451	23.8	76.2	4.81	.03*
switched	577	18.2	81.8		
natural left	168	22.2	77.8	4.03	.05*
switched left	327	16.3	83.7		
natural right	283	24.7	75.3	2.93	.09
switched right	250	20.8	79.2		

crosses with natural laterality were about 5.6 % more successful than those played with switched laterality. This effect was the same on both sides, although it only had a weak statistical significance on the right-hand side (Tab. 4.4).

The INTERRUPTION TIME for free kicks was on average 46.0 ± 14.9 s. A moderate correlation with the position of execution was found (NSR = .46). Average interruption times of around 65 s occurred in the central zones next to the edge of the penalty area, while these times were between roughly 35 and 45 s on the pitch peripheries (Fig. 4.5a). Analogously, PLAYERS IN WALL decreased with a decrease in centrality and distance (NSR = .01) (Fig. 4.5b). Roughly six players took part in wall formations in the central area, whereas on the periphery, only 1-2 players did. Correspondingly, for shots at goal from the center, walls were made up of 3.8 ± 1.7 players, and 1.4 ± 0.6 players for crosses.

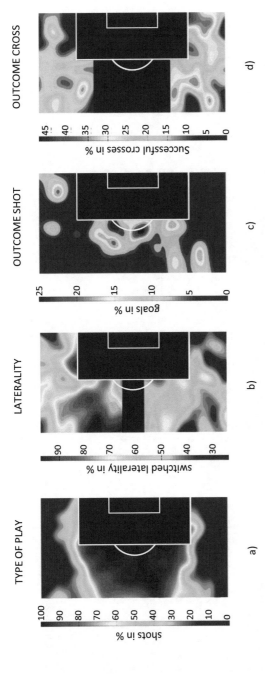

Fig 4.4: Spatial distribution of free kick performance variables: (a) proportion of free kicks with type of play = shot at goal (n = 1,833), (b) proportion of free kicks with LATERALITY = switched (n = 1,820), (c) proportion of free kicks with OUTCOME SHOT = goal (n = 968) and (d) proportion of free kicks with OUTCOME CROSS = successful (n = 1,028). The central zones in maps b and d were excluded by definition.

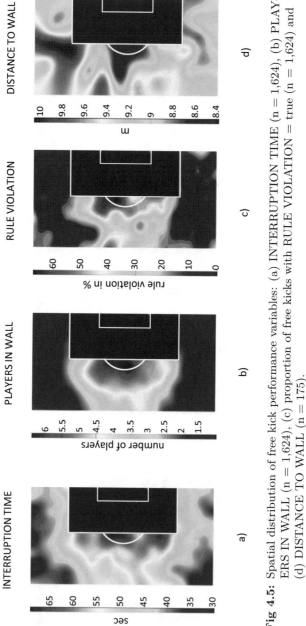

Fig 4.5: Spatial distribution of free kick performance variables: (a) INTERRUPTION TIME (n = 1,624), (b) PLAYERS IN WALL (n = 1,624), (c) proportion of free kicks with RULE VIOLATION = true (n = 1,624) and (d) DISTANCE TO WALL (n = 175).

A RULE VIOLATION was observed for 17 % of all free kicks. The variography showed a strong correlation with the location (NSR = .22). The probability was about 10 % for the peripheral areas, whereas there was an observable reduction in distance by the attacking players in the goal critical zones for approximately 50-60 % of the free kicks (Fig. 4.5c). A comparison of shots with the wall committing rule violations (n = 51) vs. no rule violation (n = 48), with the wall being in a central position, and situated less than 26.5 m away from the goal, showed no significant differences success-wise ($\chi^2 = .52, P > .77$).

The DISTANCE TO WALL was on average 9.24 ± 0.90 m. The deviation from the nominal value was 0.66 ± 0.60 m. Although geostatistically there was a strong spatial dependence (NSR = .23), the value pattern was not as conclusive. The prescribed wall distance of 9.15 m was usually undercut during free kicks near the penalty area and was exceeded for long distance free kicks or free kicks taken from the sides (Fig. 4.5d). The comparison of free kicks during which the regular wall distance was 0.5 m shorter (n = 41) with the group in which it was 0.5 m longer (n = 49) showed no differences with regard to success of shots ($\chi^2 = .66, P > .50$) and crosses ($\chi^2 = .57, P > .45$).

4.4 Discussion

The aim of this study was to describe the relationship between the location of execution, and the variables that characterize a free kick, as well as to analyze these with regard to performance. The ISO-map approach uses an averaging process and was based on moderate smoothing owing to the large variation of the performance indicators. Thus, the ISO-maps do not represent exact variable values; rather, they provide a visualization of the spatial change in value ranges. The statistical analysis was carried out on the basis of the parameters χ^2 and NSR. While the χ^2-test compares data of groups of free kicks from different discrete spatial zones with each other, variography compares data of free kicks which are within a certain distance range from each other pairwise. NSR represents the spatial dependency over the whole defensive third, while the χ^2 statistics examines differences between discrete pitch zones within this third. ISO-maps, variography and group statistics therefore shed light on different aspects of the spatial dependency of the parameter values so that seemingly divergent results (such as the spatial dependency of the success of crosses) are not necessarily conflicting results.

In total, 21,414 free kicks were taken during the 2013/14 and 2014/15 German Bundesliga games. This means that there were 34.9 free kicks per game on average, which is in line with previous studies (Casal et al., 2014; Ensum et al., 2000; Olsen et al., 1995). The increased incidence of free kicks in the defending team's central penalty area can be explained by the great number of fouls committed by the attacking team (e.g., in the context of challenge for the ball). The four widthwise strips may be derived from the common four-player-chain-formations of many teams. In these areas, there might be an increased probability of players being present, and, therefore, an increased probability of fouls occurring. One hypothesis to explain the tendency for free kicks to occur at the sidelines could be the heightened frequency of tackling. Defending players might have expected a better risk reward relationship for tackling. This is because when successful, the ball can, in contrast to when tackling occurs in the middle of the field, be played so that it is no longer on the pitch, thus ensuring an interruption to the game. The decreasing number of free kicks occurring with decreasing distance to the opponents' goal can be explained by the fact that these zones cannot be reached as often (Link, 2014). Another reason could be that fouls are risked less when defending players are closer to their own goal. This can be explained by the weighting of the danger of a goal resulting from a possible free kick against the consequences of losing a challenge for the ball (Johnson, 2006).

Only the 5.8 free kicks per game which were closer than 35 m to the goal should be directly relevant for goals. As Fig. 4.5a shows, centrality of the free kick has a crucial influence on the decision to take a shot. Players seem to view the angle to the goal as more important for success than distance. The relatively low proportion, 38.3 %, of the free kicks that were played as crosses points up the fact that players do not evaluate the chances of success of crosses as particularly high. This seems plausible as 70.5 % of crosses are intercepted by the opponent, and only 15.5 % (21.8 % in Casal et al. study in 2014) result in shots on goal. Other studies report scoring rates of 1.1 % for crosses during open play (Vecer, 2014) and 2.3 % from indirect free kicks respectively (Casal et al., 2014).

With decreasing distance from the goal, the chance of scoring increases. This was to be expected and is consistent with the results presented in a study by Pollard et al. (2004). Potentially, this can be explained by the shorter amount of time that the goalkeeper has to react and/or a higher target accuracy (Alcock, 2010; López-Botella & Palao, 2007; Morya et al., 2005). In answer to the question as to which of these effects is most apparent, one could argue that a significantly reduced target accuracy for free kicks taken from further away also means more shots that miss. As this cannot be observed, the amount of time available to the goalkeeper may be hypothesized as the more important factor.

The high proportion of shots with switched laterality indicates that this variant is more likely to be categorized as promising regarding success. Although the differences in this study were not significant, the slightly higher scoring rate (+4.3 %) could indicate that this expectation is not completely unjustified. The higher proportion of misses (+8.1 %) could perhaps be interpreted as demonstrating a higher risk for shots.

For crosses, no such clear differences in LATERALITY were found. They were usually executed with switched laterality on the left side, and natural laterality on the right. This could be owing to the higher proportion of right-footed players in the population. Natural laterality increases the probability of success (+5.6 %) of a cross. An explanation for this could be that this kind of execution type favors an out-swinging trajectory of the ball. The ball thus tends to turn away from the goalkeeper as well as the defending players (as long as they are closer to the goal and are not playing the offside trap), and so possibly increases the probability of a player from the attacking team making ball contact. This is consistent with the results of a corner kick study by Taylor et al. (2005), who, albeit based on a small sample consisting of 145 corners, determined that there were 14 % more shots on goal if the ball trajectory was out-swinging.

Crosses also show a 10.0 % higher probability of success if they were played from the right-hand side. Two explanations can be put forward for this phenomenon: Firstly, more free kicks were executed from the right side with natural laterality, which results in an out-swinging trajectory. Secondly, everyone has a dominant eye that processes and transmits information better (Knudsona & Klukab, 1997). When a free kick is played from the right side, the ball usually flies towards the attacking players from the right side and towards the defending players from the left side (because of their orientation towards or with their back to the goal). Since more people are right eye dominant (Bourassa, McManus & Bryden, 1996), it might be easier for the attacking players to anticipate and to intercept these free kicks.

These last two results are not consistent with those of Casal et al. (2014), who did not find any differences in the proportion of shots on goal or goals in terms of execution side and LATERALITY. This could be owing to the differently chosen success criteria. In contrast to previous studies (Casal et al., 2014; Taylor et al., 2005; Vecer, 2014), where the ratio of goals and/or shots on goal were used as criteria for success, we also included a controlled flick-on by the attacking team in this category. In our interpretation, in the case of a flick-on there is a realistic chance of a dangerous situation being created, even though no goal was scored, or no shot was taken. Independently, a study of corner kicks (Casal, Maneiro, Ardá, Losada & Rial, 2015) also reported an increase of 7.3 % in the proportion of shots on goal when executed from the right and an increase of 8.1 % for natural laterality. Whether both these factors benefit scoring rate could be reliably verified in this study. However, it might be a good idea to investigate whether playing the majority of crosses with a switched laterality (60 % in this study) is advisable from a tactical viewpoint. Regarding INTERRUPTION TIME, PLAYERS IN WALL and RULE VIOLATIONS, the respective ISO-maps show a radially decreasing profile of the parameter values from the middle point of the 16 m line. This can be interpreted as a tactical reaction to the increasing probability of success for central free kicks that are played from positions close to the penalty area. It could be assumed that the probability of scoring from shots on goal also decreases radially (Pollard et al., 2004). However, this cannot be conclusively verified due to the low number of goals (n = 74). The generated ISO-maps for PLAYERS IN WALL can be used as an indicator for how many players should be included in the formation of the wall. The increase of INTERRUPTION TIME near the goal is consistent with Siegle and Lames (2012). Possible reasons could be longer interruption periods due to injuries to the fouled player, the player committing the foul being penalized, the formation of a wall made up of several players, corrections to the wall formation by the referee or a lengthened period of concentration and orchestration of the player about to shoot. Further potential influencing factors include the increased use of the vanishing spray near the goal (Kolbinger & Link, 2016), as well as the score (Siegle & Lames, 2012). For this reason, free kicks where vanishing spray was used were excluded from this sample.

The wall distance of 9.15 m prescribed by the rulebook (IFAB, 2016) is estimated by referees during matches. The estimates usually have an error of about 0.60 m. The ISO-map patterns of DISTANCE TO WALL can be interpreted as a weak indication that the players shorten the distance during goal-critical free kicks when forming the wall. Furthermore, the somewhat greater wall distances during free kicks from the sides and free kicks from positions that are far away from the goal indicate that players and referees do not see distance as a relevant factor here.

The data shows that neither DISTANCE TO WALL nor RULE VIOLATION has an effect on the success of free kicks. Thus, the advantage of a smaller distance between the ball and the wall within the possible limits is too small to generate a crucial advantage for the defending team. If there is one, then it might only be relevant in specific cases. Owing to the small sample, the results discussed in this paragraph must be viewed with reservation.

The variography shows that the individual performance indicators are influenced differently by the position of execution. The weak spatial dependence of OUTCOME OF CROSS and OUTCOME OF SHOT compared to PLAYERS IN WALL, RULE VIOLATION, TYPE OF PLAY, DISTANCE TO WALL and INTERRUPTION TIME can be explained by the fact that they are not solely determined by the location of execution. They are more heavily influenced by individual constellations, group dynamics, fast accelerations, position changes, and random factors, which create perturbations in the defensive system (Hughes et al., 1998; James et al., 2012; Lames & McGarry, 2007). Individual skills, such as good timing ability or jumping power are also crucial in determining which player is the first to make ball contact after a cross. For OUTCOME OF SHOT, it must be considered that goals only occur rarely, and that the variography only reveals a limited significance in this case. Aside from this consideration, all performance parameters demonstrate a nugget value greater than 0. This means that variability cannot be explained solely by the location for any parameter. Rather, a whole range of potential influencing factors come into play, such as the score, the remaining playing time or game tactical elements, such as coach's instructions or the assessment of the abilities of the player executing the shot.

4.5 Conclusion

Geostatistical approaches can be used to examine the continuous spatial profiles of free kick parameters in soccer. According to the results of this study, crosses played into the penalty area are not very likely to result in scoring. It might be more effective to increase the proportion of passes from side free kicks and to try to reach the opposing team's penalty area using short passes and dribbles. As natural laterality promises a higher probability of success, this should also be accorded particular importance on the left side of the pitch. Since right-side free kicks tend to be more successful, it might be worthwhile practicing the defense of balls that come in from this side.

Acknowledgments   This chapter is taken from Link, D., Kolbinger, O., Weber, H. & Stöckl, M. (2016). A topography of free kicks in soccer. *Journal of Sports Sciences, 34*(24), 2312–2320. doi:10.1080/02640414.2016.1232487.

5 Match Importance Affects Player Activity

Abstract This research explores the influence of match importance on player activity in professional soccer. Therefore, we used an observational approach and analyzed 1,211 matches of German Bundesliga and 2nd Bundesliga. The importance measurement employed is based on post season consequences of teams involved in a match. This means, if a match result could potentially influence the final rank, and this rank would lead to different consequences for a team, such as qualification for Champions League opposed to qualification for Europe League, then this match is classified as important; otherwise not. Activity was quantified by TOTAL DISTANCE COVERED, SPRINTS, FAST RUNS, DUELS, FOULS and ATTEMPTS. Running parameters were recorded using a semi-automatic optical tracking system, while technical variables were collected by professional data loggers. Based on our importance classification, low important matches occurred at the beginning of round 29. A two-way ANOVA indicates significantly increased FAST RUNS (+4 %, d = 0.3), DUELS (+16 %, d = 1.0) and FOULS (+36 %, d = 1.2) in important matches compared to low important ones. For FAST RUNS and FOULS, this effect only exists in Bundesliga. A comparison of the two leagues show that TOTAL DISTANCE COVERED (+3 %, d = 0.9), SPRINTS (+25 %, d = 1.4) and FAST RUNS (+15 %, d = 1.4) are higher compared to 2nd Bundesliga, whilst FOULS is less in Bundesliga (-7 %, d = 0.3). No difference in player activity was found between matches at the beginning of a season (round 1-6) and at the end of a season (round 29-34). We conclude that match importance influences player activity in German professional soccer. The most reasonable explanation is a conscious or unconscious pacing strategy, motivated by preserving abilities or preventing injury. Since this tendency mainly exists in Bundesliga, this may suggest that more skilled players show a higher awareness for the need of pacing.

5.1 Introduction

Physical activity in soccer is subject to fluctuation. Research demonstrates that the total running distance declines after the most intensive five-minute interval (Varley, Elias & Aughey, 2012), and also from the first 15 minutes of the first half compared to the second half (Carling & Dupont, 2011). Other studies found a decline of high-intensity running from the first to the second half (Di Salvo et al., 2013). Carling and Bloomfield (2010) found that early suspensions during a match lead to an increase of total distance covered, but no change in high intensive running and technical parameters. Harper et al. (2014) report a lower passing and tackling ratio in the extra time of a match.

The reasons for variations are highly complex. One explanation for decreased running activity in the second half or after high intensity periods could be fatigue, which can be caused by physiological factors such as reduced level of creatine phosphate (Mohr, Krustrup & Bangsbo, 2005) or exhausted muscle glycogen stores (Bendiksen et al., 2012). Others suggest that player activity is influenced by tactical considerations, such as the score line or the introduction of substitutes (Bradley & Noakes, 2013). Alternatively, a reduced running performance can be attributed to conscious or subconscious pacing strategies that intend to preserve the ability to undertake high intensity activities when they are necessary (Drust, Atkinson & Reilly, 2007; Edwards & Noakes, 2009; Paul et al., 2015).

Variations in performance do not only exist during single matches but also throughout a typical season (Gregson, Drust, Atkinson & Salvo, 2010; Rampinini, Coutts, Castagna, Sassi & Impellizzeri, 2007; Silva, Magalhães, Ascensão, Seabra & Rebelo, 2013), and can be influenced by team formation, team quality, opponent strength, location or temperature (Bloomfield et al., 2005; Bradley et al., 2011; Castellano, Blanco-Villaseñor & Alvarez, 2011; Lago-Peñas & Lago-Ballesteros, 2011; Link & Weber, 2017). Another important factor – which is probably a key motivation for pacing – is the importance of a match. Particular matches at the end of a season may be more or less important to a team with respect to a reachable final rank.

Up to the knowledge of the authors, the only study which includes this factor was published by Bradley and Noakes (2013). The authors could not find any effect of match importance on total distance covered, but they reported a significant drop in high intensity running in the second half of critical matches. This decline was also found by Carling and Dupont (2011), but without including the match importance factor, so it is unclear if the drop can be explained by the criticality of a contest. However, the study is limited to a small sample size of 55 players and details of the importance measure are vague, with the studying focusing on alternative factors.

The aim of this study is to assess the importance of matches throughout two seasons of Germany's professional soccer leagues to study player activity with regard to match importance and league (skill level). For the evaluation we used physical (running distance, high intensive runs) and technical indicators (fouls, duels, attempts), from which we believe could be sensitive to pacing. We hypothesized that the activities in less important matches would decline, since players try to control their physical effort by reducing running, as well as their risk of injuries or suspensions by playing less aggressive in competitions for possession of the ball.

5.2 Methods

5.2.1 Subjects

In line with the objectives of our study, we apply a non-participative observational approach. The sample comprises 1,222 matches of Germans professional soccer leagues, Bundesliga and 2nd Bundesliga, during the 2011/12 and 2012/13 seasons. Activity data was collected for players that were on the field for at least 45 minutes. The data is publicly available on the official website of the Bundesliga so the usual approval from the research ethics committee was not required.

5.2.2 Activity Parameters

For each match, player activity parameters were collected according to Tab. 5.1. Running parameters were recorded in 25 Hz using a semi-automatic optical tracking system (VISTRACK, by Impire Corp., Germany). Technical data was observed by professional data loggers based on video recordings. All data was collected by behalf of the German Professional Soccer League (DFL). The reliability of manual data collection was secured by DFL. Complete definitions, derivation of thresholds, validation procedures and results can be found here (Deutsche Fußball Liga, 2014). The validity and reliability of the tracking system have been described by Siegle, Stevens and Lames (2013).

Tab. 5.1: Variables and operationalization. Variables represent data for one
team in one match.

ID	Definition
TDC	Average distance covered (in m) by all field players. Game stoppages were included.
SPRINTS	Number of sprints. A sprint is a time period in which players speed is (1) higher than $4.0\,\mathrm{m\,s}^{-1}$ for at least 2 seconds, and (2), within these 2 seconds, reaches more than $6.3\,\mathrm{m\,s}^{-1}$ for at least 1 second.
FAST RUNS	Number of fast runs. A fast run is a time period that fulfils the criteria (1) of a sprint but not the criteria (2) but speed reaches more than $5.0\,\mathrm{m\,s}^{-1}$ for at least 1 second.
DUELS	Number of duels. A game action is called a duel if two player of different teams are in competition for the ball. A duel is always count for both teams.
FOULS	Number of fouls. A foul is a situation, where the referee interrupted the match cause of rough play; hand ball is not included.
ATTEMPTS	Number of attempts to score a goal. We do not differentiate between shots on target or shots next to or above the goal.
LEAGUE	Bundesliga (1st)
	2nd Bundesliga (2nd)
PERIOD	Round 1-6
	Round 29-36
IMPORTANCE	Important (I): miPSC (match, team) > .05 and miPSC (match, opponent) > .05
	Medium important (M), subtype important (M-I): miPSC (match, team) > .05 and miPSC (match, opponent) < .05)
	Medium important (M), subtype low important (M-L): miPSC (match, team) < .05 and miPSC (match, opponent) > .05
	Low important (L): miPSC (match, team) < .05 and miPSC (match, opponent) < .05

5.2.3 Match Importance

To classify matches for teams as being important or less important, we apply
a further development of an existing measure of match importance. The
original measure, conceived by Bedford and Schembri (2006), is based on a

definition described in studies by Morris (1977) and Schilling (1994), where match importance is defined as the difference between the probability of a team achieving a season outcome given they win their next match; minus the probability of the team achieving the same outcome given they lose their next match. Therefore, our approach bases on the following concepts:

- *PSC*: post season consequence (PSC) is a set of positions that lead to equal consequences for a team after the season. As Tab. 5.2 shows, in the 2012/13 Bundesliga season there were 7 of these groups (e.g. PCS C = Bundesliga Champion, PCS CL = Champions League, PCS CLQ = Champions League Qualification, etc.). These may vary from season to season.

- *pPSC X (r, t)*: probability for a team t of finishing the season in PSC X in round r. This is calculated by using a cumulative binomial distribution function, which takes the points of t reached before round r and the estimated points necessary to reach a position in PSC X as an input. For example, a team on position 5 after round 30 might have a probability of *pPSC C(30, t) = .02* to become Bundesliga champion and a probability of *pPSC CL (30, t) = .23* to reach Champions League. Details of the method can be found here (Bedford & Schembri, 2006).

- *iPSC X (m, t)*: importance of a match m for a team t reaching PSC X. This is calculated by the difference between pPSC X after this round, given they win their next match; minus the pPSC X given they lose their next match. For example, a team on position 3 after round 31 might have a chance of becoming Bundesliga champion of *pPSCC = .02* when they lose the next match, and of *pPSC C = .22* if the win the next match. The importance of the match for t for becoming Bundesliga champion is *iPSC C (m, t) = .20*.

- *miPSC (m, t)*: maximum of all iPSC X of a match m for a team t.

The miPSC metric is used to define groups of IMPORTANCE (Tab. 5.1). A match m with teams A and B involved, is called *important for team A*, if *miPSC (m, A) > .05* and *miPSC (m, B) > .05*. In other words, if winning or losing this match effects the probability of achieving a different the end-of-season consequence by more than 5 % for both teams, then the match is important. In *low important* matches, *miPSC* do not exceed .05 for each of the teams, which means that there is no realistic chance to achieve a better or worse position. If *miPSC* is greater than .05 for only one team, than the match is classified as medium important. Here we also use two subgroups: If *miPSC* is greater than .05 only for team A, then this match as classified as *medium important*, subtype I for Team A, and as medium important, subtype L for Team B.

An example showing the match importance is given in Tab. 5.2. The match between FC Schalke 04 (S04) and SC Freiburg (SCF)is classified as Important, because a win may secure each team a place in the Champions League. It can also be seen that the match 1. FC Nürnberg (FCN) against SV Werder Bremen (SVW) is Low important. This is because their standings and the standing of both teams has already been mathematically assured. The match Borussia Dortmund (BVB) against 1899 Hoffenheim (TSG) is Medium important. Although the match cannot affect the standing of Dortmund (subtype L), Hoffenheim could climb to position 16 if they win and simultaneously FC Augsburg (FCA) would lose against SpVgg Greuther Fürth (GRF), which means that they will qualify for relegation play offs (subtype I).

5.2.4 Statistical Analysis

The statistical analysis uses the data of one match and one team as a statistical unit. For our analysis, we group these units according to the conditions IMPORTANCE, LEAGUE and PERIOD (Tab. 5.1). The data is presented as the mean ± standard deviation. Before using parametric statistical test procedures, the assumptions of normality were verified. A two-way (2x3 design) analysis of variance (ANOVA) was tested on each dependent variable to examine the effect of IMPORTANCE and LEAGUE condition on player activity. Also, a one-way ANOVA was used to test the effect of PERIOD on activity. Differences of activity indicators between groups were determined by pair-wise Bonferroni post hoc analysis. Effect size was calculated according to Cohen by between group means divided by the standard deviation of both groups. Tests were conducted using 95 % confidence (alpha level of .05). To control for type I error, a Bonferroni adjustment was applied by dividing the alpha level by the number of dependent variables. All statistical analyses were conducted using SPSS Statistics 22 for Windows (by IBM Corp., USA).

Tab. 5.2: Ranking, post season consequences (PSC), schedule and match importance in Bundesliga 2012/13. There are nine PSC groups (C = Bundesliga Champion, CL = Champions League, CLQ = Champions League Qualification, EL = Europa League, ELQ1 = Europa League Qualification Round 1, ELQ1 = Europa League Qualification Round 2, M = Midfield RP = Relegation Play Off, R = Relegation). Position (Pos) shows the current ranking of a team and Position opponent (PosO) the ranking of a team's next opponent after round 33. The miPSC' columns contain the number 1, if the value is greater than .05 for the team (T)and its opponent (O) and 0 otervise. IMPORTANCE (IMP) shows the classification of the matches in in important (I), medium important, subtype important (M-I), medium important, subtype low important (M-L) and low important matches (L).

Ranking after round 33				Match importance in round 34			
Pos	Team	Points	Goaldiff	PSC	miPSC' T	miPSC' O	PosO	IMP
1	FCB	88	+79	C	0	1	8	M-L
2	BVB	66	+40	CL	0	1	17	M-L
3	LEV	62	+25	CL	0	1	7	M-L
4	S04	52	+7	CLQ	1	1	5	I
5	SCF	51	+6	EL	1	1	4	I
6	SGE	50	+3	ELQ1	1	0	10	M-I
7	HSV	48	-10	ELQ2	1	0	3	M-I
8	BMG	47	-3	M	1	0	1	M-I
9	H96	42	-5	M	0	0	15	L
10	WOB	42	-5	M	0	1	6	M-L
11	VFB	42	-18	M	0	0	12	L
12	M05	41	-2	M	0	0	11	L
13	FCN	41	-9	M	0	0	14	L
14	SVW	34	-15	M	0	0	13	L
15	DUS	30	-15	M	0	0	9	L
16	FCA	30	-20	RP	1	0	18	M-I
17	HOF	28	-26	R	1	0	2	M-I
18	GRF	21	-32	R	0	1	16	M-L

5.3 Results

Our results include data from 1,222 matches. Due to poor data quality, e.g. heavy fog on the pitch, 11 matches were omitted, resulting in a final total of 1,211 matches. Based on our IMPORTANCE classification, medium important and low important matches occurred at the beginning of round 29. Up until the final round, the quantity of these matches increase due to the decrease of points that can be collected by teams (Fig. 5.1).

Tab. 5.3 shows the results from the match statistics for important, medium important and low important matches. A two-way ANOVA indicates a significant main effect of IMPORTANCE on FAST RUNS, DUELS and FOULS. In important matches, the number of FAST RUNS and DUELS were higher compared to medium important (FAST RUNS: +6 %, d = 0.5; DUELS: +9 %, d = 0.6) and low important matches (FAST RUNS: +4 %, d = 0.3; DUELS: +16 %, d = 0.9). FOULS in important matches were only significantly increased compared to low important matches (+36 %, d = 0.9). No differences were found for TDC, SPRINTS and ATTEMPTS. Also no differences were observed between the subtypes of medium important matches.

Tab. 5.4 shows the results differentiated between skill levels of teams. A significant main effect of LEAGUE was found for TDC ($F_{(1, 2,421)} = 28.5, p <$ .001), SPRINTS ($F_{(1, 2,421)} = 62.2, p < .001$), FAST RUNS ($F_{(1, 2,421)} = 46.7, p < .001$), and FOULS ($F_{(1, 2,421)} = 11.0, p < .001$). In Bundesliga, TDC (+3 %, d = 0.9), SPRINTS (+25 %, d = 1.4) and FAST RUNS (+15 %, d = 1.4) are higher compared to 2nd Bundesliga, whilst FOULS is

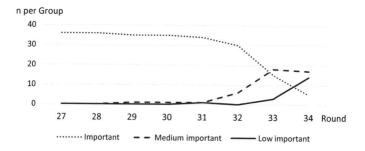

Fig 5.1: Group sizes of important, medium important and low important matches during the course of a season.

Tab. 5.3: Activity indicators grouped by IMPORTANCE. *significant differences between groups ($\alpha < .05$), ^{n} indicates significant differences (post hoc analysis) to group n

	I	M	L	F	p$<$	M-I	M-L	F	p$<$
N	2,298	88	36			44	44		
TDC	113.9 ± 4.9	112.3 ± 4.4	113.0 ± 4.8	1.8	.19	112.8 ± 5.4	112.3 ± 4.7	0.1	.69
SPRINTS	146.3 ± 27.4	146.8 ± 28.0	149.0 ± 30.2	0.1	.91	145.7 ± 27.2	148.0 ± 34.3	0.1	.75
FAST RUNS	362.7 ± 36.1^{ML}	345.4 ± 31.5	349.9 ± 40.8	3.6	.03*	347.5 ± 47.6	343.4 ± 46.0	0.4	.69
DUELS	231.2 ± 33.2^{ML}	221.2 ± 34.5	198.9 ± 35.7	16.0	.01*	113.9 ± 19.2	114.3 ± 17.0	0.1	.92
FOULS	15.5 ± 4.3^{L}	19.0 ± 4.6	11.4 ± 5.1	6.2	.01*	14.6 ± 4.9	15.9 ± 4.6	1.5	.23
ATTEMPTS	13.0 ± 4.8	13.6 ± 4.9	13.0 ± 4.7	0.7	.52	13.8 ± 4.9	13.3 ± 4.9	0.3	.71

Tab. 5.4: Activity indicators grouped by IMPORTANCE and LEAGUE. *significant differences between groups ($\alpha < .05$), n indicates significant differences (post hoc analysis) to group n

LEAGUE		All	I	M	L	F	p <
N	1st	1,206	1,144	48	14		
	2nd	1,216	1,154	40	22		
TDC	1st	115.1 ± 4.3	116.1 ± 4.2	114.4 ± 3.9	116.0 ± 4.6	1.8	.18
	2nd	111.3 ± 4.4	111.7 ± 4.5	110.4 ± 4.1	111.1 ± 4.0	0.9	.42
SPRINTS	1st	161.8 ± 22.1	161.8 ± 21.3	165.3 ± 15.2	167.8 ± 27.5	0.6	.57
	2nd	129.4 ± 24.1	130.9 ± 24.0	124.8 ± 23.5	137.0 ± 26.3	1.0	.38
FAST RUNS	1st	385.1 ± 37.2	390.5 ± 36.1^M	367.0 ± 31.5	375.3 ± 40.8	5.4	.006*
	2nd	333.7 ± 37.8	335.2 ± 37.5	319.5 ± 40.0	333.7 ± 36.2	1.6	.19
DUELS	1st	229.5 ± 34.5	230.6 ± 33.8^{ML}	206.7 ± 33.4	205.9 ± 38.1	11.1	.002*
	2nd	230.7 ± 32.1	231.9 ± 32.5^L	216.7 ± 28.0	205.9 ± 34.2	5.6	.005*
FOULS	1st	15.0 ± 4.3	15.4 ± 4.4^L	13.8 ± 4.9	11.3 ± 5.3	7.3	.001*
	2nd	16.0 ± 5.9	16.4 ± 5.5	15.9 ± 6.0	15.0 ± 2.4	1.1	.36
ATTEMPTS	1st	13.1 ± 5.0	13.0 ± 4.8	13.8 ± 4.7	12.5 ± 6.2	1.2	.30
	2nd	13.2 ± 4.6	13.1 ± 5.2	13.2 ± 5.2	13.5 ± 6.2	0.1	.89

Tab. 5.5: Activity indicators in matches grouped by PERIOD.

	Round 1-6	Round 29-34	F	p <
N	428	432		
TDC	110.8 ± 4.5	109.4 ± 4.6	0.1	.81
SPRINTS	146.1 ± 27.3	146.9 ± 29.2	0.1	.88
FAST RUNS	363.6 ± 32.5	362.1 ± 33.8	0.1	.91
DUELS	235.2 ± 34.9	233.2 ± 33.4	0.2	.78
FOULS	16.0 ± 4.4	15.6 ± 4.8	1.7	.19
ATTEMPTS	13.0 ± 4.8	13.2 ± 4.9	0.4	.53

less in Bundesliga (-7 %, d = 0.2). No significant differences were found for DUELS and ATTEMPTS.

Significant LEAGUE × IMPORTANCE interactions were found for involvement with FAST RUNS ($F_{(2, 2,421)} = 3.1, p \leq .05$) and FOULS ($F_{(2, 2.421)} = 4.8, p < .05$). In Bundesliga, FAST RUNS increased from medium important to important matches (+6 %, d = 0.6). No significant change of this parameter was found between the importance groups in 2nd Bundesliga. Also, in Bundesliga we observed more FOULS in important matches compared to low important matches (+36 %, d = 0.9). In 2nd Bundesliga, this effect did not exist. No significant interactions were found for the other dependent variables.

Tab. 5.5 shows the activity data for the groups of matches in the beginning and in the end of a season. No differences in player activity were found between these groups.

5.4 Discussion

The aim of this study was to identify influences of match importance on performance indicators in two seasons of German Professional soccer. In this context, the study employs a probabilistic model of match importance and applies the resulting classification on an analysis of activity.

Our classification method bases on the probability of a match to affect post-season consequences. With this, only the last games of the season are possibly classified as unimportant. Whilst this is one method, other techniques may also use different metrics. For example, matches against teams close in the table or below should be 'must-win' games; and matches against opponents that are higher up in the table might be considered as less important, since they are very likely to lose anyway. This means, also in earlier stages of the season, before everything is decided, unimportant matches due to this criterion could occur. On the other hand, matches might be important even when they have no consequences on the final rank, e.g. when local rivals are playing against each other. One could also argue that players should be motivated to win every game; be it for the fans or teammates, for their own honor, or for the match bonuses. However, that may be, the applied importance quantity should provide a valid model for one central aspect of match importance.

To describe high intensity activities, we used the concept of fast runs and sprints. By using the number of these events instead of distances or time spent in speed intervals, we want to accent the acceleration component of running activity. We believe this is more promising to study pacing effects, since this factor is important for describing the energy costs in soccer (Osgnach, Poser, Bernardini, Rinaldo & Di Prampero, 2010; Di Prampero et al., 2005).

The statistical analysis showed a medium effect of IMPORTANCE on DUELS in both leagues. This suggests that players are being more aggressive and play with a higher engagement due to the match having more meaning to it. IMPORTANCE also affects FAST RUNS and FOULS, but only in Bundesliga. This may suggest that more skilled players show a higher awareness for the need of pacing. This finding is consistent with (Ely, Martin, Cheuvront & Montain, 2008) which report that high level marathon runners show more adaptions to the environmental temperature compared to mid-level runners.

In order to explain the different findings for SPRINTS (not increased) and FAST RUNS (increased), we qualitatively analyzed around 50 of these events based on video recordings. In our sample sprints are often committed in situations, when there is a fast counter attack, or the chance for an own scoring opportunity. These situations require maximum effort regardless if the match is important or not. In contrast to this, we observed that fast runs often occur in situations, where a player has to keep the team formation or move himself into a better position to receive a pass. However, in low important games there might be a tendency of players to not invest effort on these types of runs, as the consequences for the match are more indirect compared to the situations where they have to sprint.

TDC and IMPORTANCE are relatively independent. This is reasonable since TDC and intensive running are quite independent of each other (Mugglestone, Morris, Saunders & Sunderland, 2013). With regard to ATTEMPTS, one could argue that importance affects this parameter, since player might reduce their effort in attacking or defending. Whatsoever, ATTEMPTS are not influenced by IMPORTANCE in our study – maybe because the effects do not exist or cancel each other. Another possible effect might be a higher activity of a team, for which a medium important match is more important – but this was not observed in our data. We believe that influence of match importance is too small to create an effect in the small sample here. Also, it might be that the intensity is more influenced by the result of the negotiated process between the two teams during the match than by considerations before the match. This is supported by findings reporting a high correlation of running parameters of both teams involved in a match (Castellano et al., 2011).

No difference in player activity was found between matches at the beginning in the end of a season. This confirms that the decline at the end of the season in low important matches is not primary caused by fatigue. However, this is somehow remarkable since the accumulation of fatigue is a well-known phenomenon in soccer (Silva et al., 2011). The data suggests that professional soccer teams are able to compensate individual fatigue up to a certain extent, e.g. by effective recovering procedures or by changing the starting line up from match to match. However this may be, there are also some other factors that could have affected match activity in our study like a reduced training load (Gabbett & Domrow, 2007), or environment conditions (Link & Weber, 2017).

Analysis of the condition LEAGUE finds more FOULS in 2nd Bundesliga. One explanation is that players might have less technical skills, e.g. a lower passing accuracy (Bradley et al., 2013), which can lead to situations where fouls are more likely. Also, TDC, SPRINTS and FAST RUNS in 2nd Bundesliga are reduced compared to Bundesliga. This stands in contrast to the findings in UKs soccer published by Di Salvo et al. (2013), who reported the opposite effect in Premier League and Championship League. We think that there are two possible explanations. Firstly, playing style in the UK and Germany is different and this influences the relationships between the first and the second division in these countries. Secondly, this could be an artefact of the different tracking systems used. Up the experience of the authors, some systems are more sensitive for the location of the cameras than others. Since the stands are less high in the second leagues, this might increase the number of tracking lost which lead to less running distance measured due to interpolation. But this is somewhat speculative, with additional studies required to spread light on this finding.

5.5 Conclusion

This study shows that professional soccer players significantly decline their match activity in low important matches at the end of the season. This effect seems primary not to be related to fatigue, but to a conscious or unconscious pacing strategy. Since this phenomenon mainly exists in Bundesliga, our data suggests that more skilled players show a higher awareness for the need of pacing. Although this may not be attractive to supporters, this strategy could be reasonable. This might help to preserve their abilities for finals or cup competitions, whilst also preventing injury or being sent off.

Acknowledgments   This chapter is taken from Link, D. & de Lorenzo, M. F. (2016). Seasonal pacing – Match importance affects activity in professional soccer. *PLOS ONE*, *11*(6), e0157127. doi:10.1371/journal.pone.01 57127.

6 Effect of Ambient Temperature on Pacing Depends on Skill Level

Abstract This study examines the influence ambient temperature has on the distances covered by players in soccer matches. For this purpose, 1,211 games from the top German professional leagues were analysed over the course of the seasons 2011/12 and 2012/13 using an optical tracking system. The data shows a) significant differences in the total distance covered (TDC, in m $\cdot$ $(10\,\text{min})^{-1}$) between the Bundesliga (M = 1,225) and 2nd Bundesliga (M = 1,201) and b) a significant decrease in TDC from NEUTRAL (-4 to 13 °C, M = 1,229) to WARM (≥ 14 °C, M = 1,217) environments. The size of the temperature effect is greater in the Bundesliga (d = .30 vs. d = .16), even though these players presumably have a higher level of fitness. This suggests that better players reduce their exertion level to a greater extent, thus preserving their ability to undertake the high intensity activities when called upon. No reduction in running performance due to COLD (≤ 5 °C) temperatures was observed.

6.1 Introduction

It is often necessary for competitions in soccer to be held under suboptimal environmental circumstances. This raises questions about the relationship between environmental variables such as temperature, humidity, rainfall or high altitude and soccer performance (Dvorak & Racinais, 2010). FIFA's announcement of Qatar as the host country for the 2022 Soccer World Cup has led to discussions in the media and the public arena. Particular attention has been on the possible detrimental effect the extreme temperatures will have on performance. There is concern that air temperatures on the soccer pitches above 40 °C may adversely affect the players' performance, the dynamics of play, and thus the attractiveness of matches.

The issue of playing soccer in cold and hot conditions has not yet been adequately studied (Blomstrand & Essén-Gustavsson, 1987; Dvorak & Racinais, 2010; Ekblom, 1986). Two recent case studies contrasted two games in hot environments (Mohr, Nybo, Grantham & Racinais, 2012; Özgünen et al., 2010). They found evidence of a decline in running performance and an increase in core temperature of the players, particularly in the second half of matches. Mohr et al. also observed decreases in total distance covered and number of sprints made (2012) and indicated a reduction in muscular endurance during post-match analysis (2013). These results are in line with findings from laboratory research that showed that elevated ambient temperature increases the rate at which fatigue in the cardiovascular system (Asmussen & Bøje, 1945; Nybo, Møller, Volianitis, Nielsen & Secher, 2002) and the central nervous system sets in (Nybo & Secher, 2004). The result is a reduction in the supply of oxygen to skeletal muscles (Hargreaves & Febbraio, 1998) that is further aggravated by coincidental dehydration (González-Alonso, Calbet & Nielsen, 1998).

At the other end of the temperature spectrum, laboratory tests have demonstrated that low temperatures negatively affect fat and glycogen metabolism (Doubt, 1991; Sink, Thomas, Araujo & Hill, 1989). This results in reduced oxygenation of skeletal muscles (Pendergast, 1988) and subsequent lowering levels of Adenosine Triphosphate (ATP) and Creatine Phosphate (CrP) concentration that promotes accelerated lactate build up (Blomstrand & Essén-Gustavsson, 1987), thereby reducing endurance and sprint capacities (Drust, Rasmussen, Mohr, Nielsen & Nybo, 2005). Carling (2011) were unable to confirm these effects in soccer, but their sample size was relatively small (n = 29 matches) with only a few matches being played at ambient temperatures below 5 °C.

Most performance drop offs due to low and high temperatures are also seen in laboratory settings or long distance running (Montain, Ely & Cheuvront, 2007), but it is still unknown the extent to which temperature influences activity patterns in sports involving intermittent bursts of speed. Existing studies in soccer rely on small sample sizes that are not able to detect real changes due to the variability of running performance in soccer (Gregson et al., 2010). To close this gap, this paper provides empirical evidence of the relationship between air temperature and running performance based on a large set of data.

6.2 Methods

6.2.1 Approach

The study observes temperature and the running performance of professional players in official competitions in the top German leagues (Bundesliga and 2nd Bundesliga). Running performance is operationalized by the TDC for players in a single match, an aspect that was identified as a key indicator of performance in many studies (Mackenzie & Cushion, 2013).

6.2.2 Subjects

The sample comprises the majority of games during the 2011/12 and 2012/13 seasons. A few matches (n = 11) with poor data quality e.g. due to heavy fog on the pitch, were omitted. The physical performance of players in a total of 1,211 matches involving 38 teams was analyzed. The sample only contains data for all players that were on the field for at least 45 minutes. This study is in line with the American College of Sports Medicine's policies regarding animal and human experimentation, as providing these data is a condition of the players' contracts of employment. The data is publicly available on official websites so the usual approval from the research ethics committee was not required. Informed consent statements were given by all subjects when signing the contract as a professional soccer player.

6.2.3 Procedures

The German Soccer League systematically records the ambient temperature (T) within about 10 minutes prior to kick-off based on official data supplied by the German Weather Service. Temperature was classified into four groups: COLD ($\leq -5\,°C$, n = 30), NEUTRAL (-4 to 13 °C, n = 1,249), WARM (14 to 27 °C, n = 428) and HOT ($\geq 28\,°C$, n = 4). As no statistically robust sample size for HOT was available, we assigned these games to the WARM group for analysis.

Running distances were recorded using a semi-automatic optical tracking system (VISTRACK, by Impire Corp.,Munich, Germany) that has a sampling frequency of 25 Hz. This system consists of two cameras that identify players by way of their movement, shape and color information. The validity and reliability of this system for taking such measurements have been described in detail elsewhere (Siegle et al., 2013). All Data was obtained from the

German Professional Soccer League (in German: DFL, Deutsche Fußball Liga) and is available to the public via its official website.

The variable total distance covered (TDC) represents the average distance covered by all field players per team per match. Due to variations in the duration of matches, all running distances were presented as the distances covered in a 10 minute period. Game stoppages were included.

6.2.4 Statistical Analyses

All statistical analyses were conducted using SPSS 22 for Windows (by IBM Corp., Armonk, NY, USA). Two-way ANOVA were used to compare performance across temperature ranges and the skill levels of the teams in order to identify interactions between these factors. Pair-wise post-hoc analysis between groups uses a one sided t-test. Effect size was calculated according to Cohen by between group means divided by the standard deviation of both groups. Normal distribution of TDC and temperature was tested using a Kolmogorov-Smirnov test. Statistical significance was set to an alpha level of .05.

For visual analysis (Fig. 6.1), matches were allocated to 39 temperature groups, each representing $1\,°C$ (from -10 to $28\,°C$). Within these groups, the average percentage deviation from the TDC mean value was calculated and filtered using a moving average of order 5 (TDC_{MA}).

6.3 Results

Fig. 6.1 illustrates TDC_{MA} across the temperature range. TDC_{MA} is mostly constant in the NEUTRAL range and decreases at lower or higher temperatures. A linear regression model using one function for each segment shows the best fit for the WARM range ($R^2 = .94$), followed by the NEUTRAL range ($R^2 = .39$) and the COLD range ($R^2 = .14$).

Fig. 6.2 shows TDC by temperature and league. A two-way ANOVA indicates a main effect of temperature ($F = 7.23$, $p < .05$). Post-hoc tests showed that this effect can be seen between NEUTRAL and WARM ($M = 1,229$ vs. $M = 1,217$; $t = 5.01$, $p < .05$, $d = .22$), as well as between COLD and WARM ($M = 1,231$ vs. $M = 1,271$; $t = 1.83$, $p < .05$, $d = .30$). Differences are not significant between COLD and NEUTRAL. Moving the threshold between NEUTRAL and WARM from 13 to $20\,°C$ enhances apparent differences ($t = 5.49$, $p < .05$, $d = .42$) between the two groups.

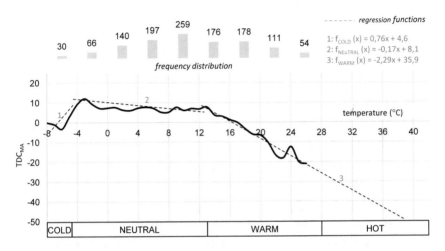

Fig 6.1: Total Distance Covered (TDC$_{MA}$) in 1,211 soccer matches across the temperature range. A reduction in the distance covered becomes evident above 14 °C. The dotted lines show three regression functions for the different temperature segments.

A second main effect is associated with league (F = 9.76, p < .05). Running distances in the Bundesliga are significantly greater than those in the 2nd Bundesliga (M = 1,225 vs. M = 1,201; t = 22.4, p < .001, d = .8). An interaction effect between league and temperature was also indicated (F = 8.13, p < .05). This can be ascribed to the decrease of TDC from NEUTRAL TO WARM, which is greater in the Bundesliga compared to 2nd Bundesliga (d = .30 vs. d = .16).

6.4 Discussion

The overall distances covered in the Bundesliga are higher than in the 2nd Bundesliga. The findings contradict results found by Di Salvo et al. (2013). They report lower distances covered in the English Premier League than in the Championship League (about 30 m per 10 min less). The differences between the two German leagues fall within the standard deviation. Hence, the interpretation by Di Salvo et al. that league differences are not relevant in practice is only partially supported. Even though expressions of physiological stress vary only slightly, these results may instead indicate

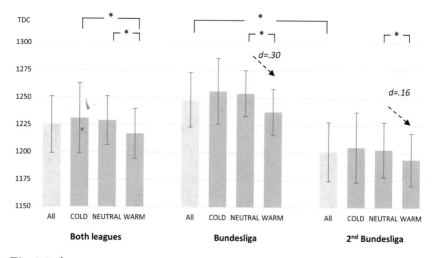

Fig 6.2:]
Total Distance Covered (TDC, in $m \cdot (10\,min)^{-1}$ per player) at different temperatures and for different skill levels. Players in the Bundesliga show significantly higher running performances ($50\,m \cdot (10\,min)^{-1}$) than players in the 2nd Bundesliga. Decreases in the distances covered in WARM environments are evident for both leagues but greater in the Bundesliga.

that opponents being attacked earlier and more intensely or offensive plays being quicker.

Running performance in the NEUTRAL environment is largely temperature independent. This supports findings from previous research on marathon races (Montain et al., 2007) that suggest ideal running conditions are found at a temperature range from 5 to 10 °C. Furthermore, our findings imply that for soccer, the NEUTRAL range extends to lower temperatures. Analysis of the COLD environments indicates that performance drops off at approximately −5 °C, but these effects are not statistically significant. It is possible that there are anomalies resulting from the small sample size. Furthermore, adverse affects may be counteracted by players taking steps to protect themselves against the cold weather. Another possibility could be that a reduction in physiological performance in soccer only becomes apparent at significantly lower temperatures. There were no data available for the observed time frame on which to assess these possibilities, however, no reliable conclusion can be drawn.

In WARM conditions, starting at approximately 14 °C, there is a notice-able reduction in the distances covered. This finding could be explained by physiological effects, such as dehydration (González-Alonso et al., 1998), or by a pacing strategy (Aughey, Goodman & McKenna, 2014). Players in the Bundesliga, who play to a higher standard, exhibit a stronger reaction to increasing temperature even though they ought to have higher levels of strength and stamina. This suggests that they are – consciously or uncon-sciously – more aware of the negative effect of warm weather and reduce the distance they cover thus preserving their ability to undertake high intensity activities when called upon. This finding is consistent with the findings of (Montain et al., 2007), which reports the same effect when comparing the winners and mid-level runners in marathon races.

No empirical conclusions can be drawn for HOT environments, but the re-gression functions can be applied in order to estimate the effects. When data is interpolated to 36 °C, a performance reduction of approximately 4 % emerges (compared to the average value in the NEUTRAL range). Subsequent performance differences are equivalent to those found when com-paring Bundesliga and 2nd Bundesliga. Even if the assumption of linearity is somewhat speculative, the findings are roughly in line with the results of Mohrs' studies (2012), where he reported reductions in total distance covered of about 7 % at 43 °C compared to 20 °C. In contrast, Özgünen et al. (2010) compared two games at 34 °C and 36 °C that showed reductions in total distance covered of about 5 %. In light of the results of the present study, such performance decreases do not appear to be typical. Further research in countries with higher average temperatures, such as Spain or Brazil, could be done in order to confirm these findings.

Interpretations need to take into account that running distances are not determined solely by physiological and pacing factors, but also by oppon-ents, playing systems, and scores, all of which significantly influence running performance (Carling, 2011; Gregson et al., 2010; Rampinini et al., 2007). Furthermore, at the beginning of a season, fatigue due to increased training intensity during pre-season preparations, European or World Champion-ships and initial formation problems affect playing behavior. In addition to this, end-of-season games may be perceived as less relevant when the teams' positions in the league table have already been established. At the same time, ambient temperatures are naturally higher at the beginning and at the end of the season in Germany. For this reason, additional studies are required to investigate possible mediators. Assuming that the above confounding variables exert no significant systematic effect due to the large sample size, the effect of air temperatures on running behavior can be confirmed.

6.5 Conclusion

The practical consequences of this fundamental research study are twofold: First, our data suggests that hot ambient temperatures cause a significant change in running activity in soccer. With a view to the 2022 World Cup in Qatar, we have to expect less dynamic games – unless preventative measures are taken. FIFA representatives would be well advised to consider countermeasures, such as installing air-conditioning in the grounds or maybe postponing competitions until the winter months (Blatter, 2013). Second, soccer coaches ought to be made aware of the effects of temperature, even in moderately warm conditions. Based on our data, we would suggest that interventionary measures are considered when the air temperature exceeds 22 °C. It might be also a good idea to sensitize players to the idea of not attempting to maintain their running performance at the same level as in colder conditions. Players should think about intelligent movement strategies that fulfil tactical objectives with a minimum of physical exertion. This could help them to preserve their ability to sprint at a level that allows them to cope with situations even at the end of the game.

Acknowledgments   This chapter is taken from Link, D. & Weber, H. (2017). Effect of ambient temperature on pacing in soccer depends on skill level. *Journal of Strength and Conditioning Research*, *31*(7), 1766–1770. doi:10.1519/JSC.0000000000001013.

7 Vanishing Spray Reduces Extent of Rule Violations

Abstract   More and more sport associations introduce innovative devices to support referees and umpires respectively, affecting a strong need for the evaluation of these devices. This study evaluates the use of the new vanishing spray for free kicks in the German Bundesliga. In more detail, the aim of the study is to investigate if the spray reduces violations of the required minimum distance and consequently the respective punishments, if it reduces errors concerning the distance set by the referee and if it leads to a higher success rate of free kicks. Therefore, 1,833 free kicks of the 2013/2014 and 2014/2015 season of the German Bundesliga were screened using a self-designed observational system. For the statistical analysis two parallel samples were built of 299 free kicks each. The results showed no decrease of free kicks with distance violations but a significantly lower extent of these violations ($\chi^2 = 4.58$; p $<$ .05). However, none of these violations were punished appropriately. Concerning the success of free kicks, no significant impact was found neither for shots nor for crosses. In addition, no influence on the distance set by the referee could be identified. The main objective of the vanishing spray was basically realized, but the use didn't lead to any further positive (side) effects. Due to the lack of punishment, the authors raise concerns about the current application of the minimum distance rule.

7.1 Introduction

An increasing amount of innovative devices is introduced in several sports to support the umpires and referees respectively. In general these devices serve as another tool to ensure the legitimate outcome of sport competitions (Rock, Als, Gibbs & Hunte, 2013). The function of the tools is to compensate the limits of human perception that cause for example optical errors and flash lag effects (Helsen, Gilis & Weston, 2006), as well as to eliminate bias that referees show towards the hosting team or players of their own race and the same country (Parsons, Sulaeman, Yates & Hamermesh,

2011; Pope & Pope, 2015; Dohmen & Sauermann, 2015). The devices can be divided up into three different groups. Devices that support the referees in their decision making, devices that replace the referees for certain decisions and devices that help them to enforce the rules of a sport.

The new vanishing spray in soccer belongs to the latter category. Referees can use this spray to mark the required minimum distance of 9.15 m (10 yd) that players of the defending team have to obey before a free kick is taken. To mark the distance the referee draws a line between the ball and the goal that players of the defending team are not allow to cross until the ball is touched by the offensive team. Its use was finally approved at the 126th annual general meeting of the International Football Association Board (IFAB) in 2012, but the spray was already introduced in several competitions across South America since 2000. The vanishing spray attracted worldwide attention with its appearance at the FIFA World Cup 2014 and was introduced in several European Competitions, like the German Bundesliga, before and during the 2014/2015 season.

Currently, there is still a lack of evaluation research for this device, representing a common issue in sports, concerning these kinds of devices. Especially the impacts of umpiring aids are neglected. In this connection, the impact is not only the achievement of the objectives but also other (side) effects of the innovation, which can be positive as well as negative. This is quiet surprising, as investigating the merit or worth of interventions, should be a main goal due to most of the common definitions of evaluation (Scriven, 1991; Stufflebeam & Shinkfield, 2007). The evaluations of the respective associations focus mostly, sometimes even exclusively, on technical parameters that the devices have to fulfill. This technical aspect has the advantage that it can be investigated under laboratory conditions and it represents the most obvious precondition for the introduction of a device. Thus, taking goal line technology as an example, the FIFA established a series of technical tests that the systems have to pass to get credited as an official device, mainly focused on the accuracy of the systems and real time detection. The same applies to scholarly studies (Psiuk, Seidl, Strauß & Bernhard, 2014; D'Orazio et al., 2009). The FIFA as well as other soccer associations did not collect data about costs and benefits of the goal line technology. Kolbinger, Linke, Link and Lames (2015) showed a very low incidence of scenes that could be resolved exclusively by goal line technology. They found less than four such incidences per season per league and therefore raised concerns about the cost-benefit ratio, especially considering the costs of round about 2.4 million per year.

For the use of the Hawk-Eye technology in Tennis, even the standard of the technical evaluations is questioned. In two articles considering the effect of the presentation of the technology on the publics' understanding of science, Collins and Evanvs (2008, 2012) denounced the test-design of the International Tennis Federation. Nevertheless some studies investigated the use and the impact respectively of this device. Mather (2008) as well as Abramitzky, Einav, Kolkowitz and Mill (2012) both found that slightly under 40 % of the challenges were successful, meaning that the technology didn't confirm the umpires' call. Abramitzky et al. (2012) also showed that it is only used for a very low share of points, almost exclusively for balls within 100 mm of the line (Mather, 2008), but that successful challenges can increase the winning probability by up to over 25 %.

No evaluation research was run for the use of the vanishing spray in soccer yet, so the aim of this study is to overcome this lack by investigating five hypotheses. In addition to those, the using patterns of the device were described. The spray was introduced to help the referee to enforce the minimum distance rule and stemming discussions about the spot of the ball or wall respectively. Thus, the first hypothesis to be tested is that the spray leads to less violations (H 1.1) – or respectively a lower extent of violations (H 1.2) – of the required minimum distance by players of the defending team. According to the official laws of the game of the Deutscher Fußball Bund (DFB, trans. German Football Association), these violations should be punished with a yellow card and a repetition of the free kick. Taken this into account, it was checked if the spray affects fewer warnings and less free kicks that have to be retaken, to provide further information concerning the patterns of rule violations. Without reference of these targets, but as the innovation created more awareness of the required distance, a third hypothesis, that the spray reduces estimation errors of the set distance between the spot of the free kick and the wall, was investigated. In addition, we assumed that no violations of the required distance benefit the kicking team, resulting in a higher success rate for free kicks taken as crosses (H 3.1) as well as for those that were taken as shots (H 3.2).

7.2 Methods

7.2.1 Data recording

The data set consisted of all free kicks of the 2013/2014 and 2014/2015 season that were taken as shots or crosses with a distance less than 35.0 m of the goal line. By signing a contract of employment as a professional soccer player in the German Bundesliga, each players sings a statement of consent to being monitored during matches. The provided data included a match ID, the teams involved, event time, two-dimensional coordinates of the spot of the free kick and whether it was taken as a shot or cross. Using a self-deigned observational system two specific trained experts collected data for further variables that are shown in Tab. 7.1. Therefore, specific videos of the free kicks were provided starting 90 s prior to the taking and ending ten seconds after it.

The distance between the spot of the ball and the wall was obtained with a custom made analysis software. Using homography, this software enables the user to determine points on the field by transforming video coordinates into real world coordinates. Therefore, this software requires not just positional data of a match, but also the respective specific tracking video. These data and videos respectively were only available to the authors for a subsample for which the distance set by the referee could be obtained. The respective variable DISTANCE ERROR shows the absolute difference to the regulatory 9.15 m.

7.2.2 Reliability

The examination of the inter-rater-reliability showed excellent scores for most of the variables. Cohen's Kappa reached a value of 1.0 for the use of the spray, .91 for the PUNISHMENT of violations and 1.0 for RESULT OF SHOT. RESULT OF CROSS and the identification of VIOLATION felt behind with .79 and .80 respectively but were still acceptable. The correlation coefficient (.98) as well as the relative observed agreement (92.6 %) for the numbers of PLAYERS IN WALL also showed a good agreement between the observers.

7.2.3 Statistical analysis

To investigate the influence of the vanishing spray, sprayed free kicks after the introduction were compared to those before, based on the idea of evaluating interventions by comparing respective variables on different time points of a program (Cronbach, 1963). Thus, the mentioned variables were collected for 1,833 free kicks in total. 1,108 of these free kicks were taken prior to the introduction of the vanishing spray and 725 after its introduction on the 8th day of the 2014/2015 season. As the free kicks after the introduction showed no consistency considering the use of the spray, it was

Tab. 7.1: The collected variables inclusive their respective categories and definitions

Variable	*Categories* and Definition
SETTING	*Spray*
	NoSpray
LOCAL CATEGORY	*Left/ Right Near*: On the sides of the penalty box
	Left/ Right Far: On the sides of the virtually extended penalty box between 16.5 m and 35 m distance of the goal line
	Central Near Left/ Right: Inside the virtually extended penalty box with not more than 26.5 m distance of the goal line (penalty box plus 10 m). Left and Right are divided by a virtual line in the middle of the field drawn at right angle to the goal line
	Central Far: Otherwise, but within 35 m distance to the goal line
PLAYERS IN WALL	Numbers of defensive players in the wall
VIOLATION	*True*: At least one player passes the referees mark with his entire foot.
	False: otherwise
MASSIVE VIOLATION	*True*: More than one player commits a violation or a player reduces the distance by more than one meter.
	False: otherwise
PUNISHMENT	*Yellow card*: A yellow card is awarded
	Verbal cautions: The referee corrects the players verbally
	None: No punishment
RETAKE	*True*
	False
SUCCESS OF SHOT	*OnTarget*: A goal is scored, the ball hits the goals border or the goalkeeper makes a save
	Missed: Ball misses the goal or is blocked by a player outside the wall
	Wall: Ball is blocked by the wall
SUCCESS OF CROSS	*Successful*: A player of the offensive team is able to perform a shot or pass with the first touch after the cross
	Not Successful: otherwise

decided to run a parallel study design to investigate the influence of the
new device. The parallelization was performed in three steps. First, the
free kicks were grouped by LOCAL CATEGORY on the basis of its two-
dimensional coordinates (Tab. 7.1). The number of PLAYERS IN WALL
served as the second criteria, representing the perceived risk of the defending team. At last, the free kicks were paired in these categories on the basis
of the shortest distance. Thus, two parallel samples were built with 299 free
kicks each (N_{Spray}/ $N_{NoSpray}$). 81 pairs of free kicks of these two groups
represented the subsample for the investigation of the set distance by the
referee.

The spatial distribution of the spray's use was visualized using the ISO-
PAR method (Stöckl, Lamb & Lames, 2011). Due to the different styles
of the obtained variables different statistical analysis were run, after veri-
fying the assumptions of normality. On the one hand, a paired t-test was
calculated for DISTANCE. On the other hand, chi square tests for VIOL-
ATION, PUNISHMENT, RETAKE and the SUCCESS of free kicks. All
statistical analysis were performed with SPSS (Version 23.0; Armonk, NY;
IBM Corp.), except the respective effect sizes Cohen's d and Cramér's V
that were calculated manually. The magnitudes of the effect sizes were
evaluated based on the limits: .10 (small), .30 (medium) and .50 (large) for
Cramér's V (Cramér, 1946). The limits for Cohen's d were .20 (small), .50
(medium) and .80 (large) (Cohen, 1992).

7.3 Results

For 308 of the 725 investigated free kicks after the introduction the referees
decided to mark the regulatory distance with the vanishing spray. Fig. 7.1
shows that the spray was used more likely for central free kicks, especially
with decreasing goal distance. The spray was used for all the investigated
respective free kicks with six or more players in the defensive wall and for
88.9 % and 89.0 % for free kicks with walls of four and five players respectively. This number decreases further for free kicks with three (70.9 %), two
(34.7 %) or one player (7.8 %) participating in the wall.

The introduction of the vanishing spray showed no significant influence on
VIOLATION (Tab. 7.2), as the share of free kicks with violations remains
on a similar level (OR = 1.02; 95 % CI: 0.69 to 1.49). However, the share
of MASSIVE VIOLATION decreases significantly by 6 %, representing a
trivial effect size but an odds ratio of 0.60 (95 % CI: 0.36 to 0.99). Despite these violations of the required minimum distance, none of the free

Tab. 7.2: Test statistics and effect sizes of the comparisons of the investigated variables before and after the introduction of the vanishing spray. Continuous variables are stated as mean ± standard deviation. (*significant differences; [A]small to medium effects)

	NoSpray	Spray			
Nominal variables			χ^2	**p**	**V**
VIOLOATION	25.4 %	25.8 %	0.01	.925	.00
MASSIVE VIOLATION	16.7 %	10.7 %	4.58*	.032	.09
SUCCESS OF CROSS	26.7 %	19.2 %	1.26	.261	.09
SUCCESS OF SHOT			0.31	.857	.03
OnTarget	34.4 %	32.3 %	0.23	.635	.02
Missed	38.8 %	41.2 %	0.26	.661	.02
Wall	26.8 %	26.5 %	0.00	.954	.00
Continuous variables			**t**		**p**
DISTANCE ERROR	0.70 ± 0.64 m	0.65 ± 0.58 m	0.56	.577	.01

kicks of the sample was retaken and no yellow cards were awarded for this reason. Regarding PUNISHMENT, six verbal cautions for such violations were recorded for the treatment group, showing a significant increase with small effect size compared to the zero verbal cautions of the control group ($\chi^2 = 6.06$; p < .02; V = .10; OR: 132.7; 95 % CI: 1.12 to $8.56 \cdot 10^3$).

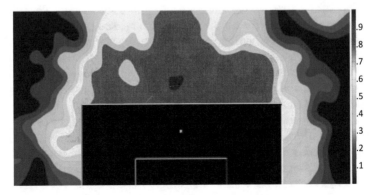

Fig 7.1: Spatial distribution of the proportion of sprayed free kicks (n = 725). Dark red illustrates a probability of 1.0 that the spray was used, dark blue a probability of 0.0.

The comparison of the set distance between the ball and the wall by the referee showed no influence of the vanishing spray in any direction. The average distance was 9.24 m for each group ($SD_{NoSpray}$ = 0.87 m, SD_{Spray} = 0.93 m). The average absolute difference from 9.15 m, the prescribed distance by the laws of the game, showed a non-significant mean difference of −5.12 cm (CI: −23.2 to 12.9). Overall, 50.6 % of the estimations are within a DISTANCE ERROR of just 0.5 m, 38.6 % showed an absolute error of 0.5 to 1.5 m and 10.8 % were off by more than 1.5 m.

The influence on the success of free kicks was evaluated separately for crosses and shots. For neither category of SUCCSESS OF SHOT, significant influences could be identified. Despite the slight increase of goals from 8.6 % to 9.3 % after the introduction, fewer shots were recorded in the OnTarget category (OR: 0.91; 95 % CI: 0.60 to 1.38) and more shots in the category Missed (OR: 1.10; 95 % CI: 0.74 to 1.65). The number of shots that were blocked by players standing in the wall decreased slightly by 0.3 % (OR: 0.99; 95 % CI: 0.63 to 1.54). SUCCESS OF CROSS was just 0.65 as high after the introduction of the vanishing spray (95 % CI: 0.29 to 1.46). As stated in Tab. 7.2, none of these changes were significant.

7.4 Discussion

The aim of this study was the evaluation of the new vanishing spray in soccer, which is used for 42.5 % of the respective free kicks, especially for those in promising positions. Therefore, five hypotheses were tested of which just one, the lower extent of rule violations, could be supported. The four other hypotheses concerning the amount of violations, the distance set by the referee and the success of either crosses or shots were not supported by the results. After separately discussing the findings regarding each hypothesis, the authors point out the significance of these findings and null-findings respectively for the evaluation research process itself as well as the understanding of the underlying phenomena.

A reduction in violations of the minimum distance, could not be proofed by the share of free kicks with violations. One out of four free kicks is still affected by such an incident. Nevertheless, the number of massive violation showed a significant decrease. Even though the respective effect size is just .09, the risk ratio shows a 1.56 times higher risk for massive violations for the control group. Thus, the spray had a positive effect on the extent of violations, and the main goal of the introduction was at least reached in curbing such incidences. In addition it is worthy to point out a limitation of this study. For sprayed free kicks, the observers could use the marked line as a reference, which (obviously) was not possible for other free kicks. Consequently, for the latter category, the observers were instructed to note only violations if they can see the defenders moving towards the ball. Thus, if the wall reduced the distance while were not covered by the broadcast, this could not be identified by the observers. This step was necessary to reach a good inter-rater agreement, but likely affected an underestimation of violations for free kicks without spray due to methodological reasons.

Despite this frequent incidence, the results also showed that violations of the distance rule were punished neither before nor after the introduction of the vanishing spray. Just a few verbal cautions were awarded after the introduction, but these verbal cautions not fulfilling the required extent of punishment. Two reasons could lead to this lack of punishment for rule violations. The first reason simply is that the referee can not identify the violation. Since a free kick is a situation that has a comparatively clear structure in which the referees can focus more specific, the authors assume that the referees should be able to identify violations of the required minimum distance. In addition, it should be even more possible for the referees with a marked line on the field, which now is available after introduction of the vanishing spray. Thus, it is possible that some kind of unwritten rules come into play. According to D'Agostino (1995), unwritten rules can be seen as unofficial, implicit conventions that determine how the rules of a game are to be applied in specific circumstances. Translated for this situation, we suspect some kind of agreement between players of the different teams and the referee that the distance rule should be enforced less rigorously for free kicks. Considering that the correct punishment of such violations is even stated one more time in an extra annotation in the official laws of the game, this could be seen as a discrepancy between the protagonists on and off the field.

For positioning the wall the referees have to estimate the distance to the ball without any external aids. In more than 50 % of the free kicks the referees were within a distance of half a meter, but also a decent amount of errors of more than 1.5 m occurred. We think the mean absolute error of 0.70 m before the introduction was still acceptable, especially considering that the mean distance to the ball is just nine cm to high. The vanishing spray can not support this process directly but led to more awareness of the distance in general. However, this did not affect the quality of estimation by the referees as there was no significant decrease in the absolute error.

Round about a third of the shots in this study were on goal, slightly more than one out of four is blocked by the wall and the remaining round about

40 % missed the target or were blocked by another player outside the wall. The results are pretty similar to previous findings of Carling et al. (2005). In an investigation of 152 attempts of the World Cup 2002 they found a rate of 8 % for goals and round about 35 % on goal (combining their respective categories). They state that only seven percent of the free kicks were blocked by the wall but a surprisingly high amount of free kicks that went through the wall (15 %), which sums up to a similar rate compared to this study.

The overall rate of successful free kicks taken as crosses was short to 23 %. To compare the values to those of previous studies, it is necessary to consider that we used a different definition of 'successful'. Casal et al. (2014) for instance stated that of 21.8 % of indirect free kicks resulted in a shot in international competitions. As mentioned above in the current study a cross was rated successful if the ball reached a player of the same team in a way that this player could execute a controlled action with the ball. That is the only goal that the player who is taken the free kick can control and vice versa for what he has to pass the defensive wall. Thus, this definition of 'successful' is more appropriate for this study and arguably also for other studies that focus rather on the taking player.

The use of vanishing spray should benefit the offensive team. Despite a slight increase of the goal rate, this could not be proved with the results of this study. No significant differences were found for the success of free kicks, neither for free kicks taken as shots nor for crosses. The rate of successful crosses even decreased by 7.5 % which, however, was not a significant difference. Comparing the free kicks with massive violations of the distance rule indicate that this violations do not affect the outcome of the free kick, but there was not a large enough sample to run reliable analysis. Another reason could be that the curbing of rule violations is too small to create a significant benefit for the kicking team, especially considering the variations in the distance set by the referee.

Summed up, the introduction of the vanishing spray basically fulfilled its main goal, by reducing the extent of violations of the minimum distance rule, but didn't lead to further positive effects. Especially for the purpose of evaluation research, the respective null findings are as valuable as other findings to estimate the worth of an intervention. Rather spoken, this worth is estimated by the respective stakeholders, which must thoroughly consider all results of an evaluation for the decision making process (e.g. Stufflebeam, 1983). This study illustrates an interesting example, as more and more competitions start to use the vanishing spray, despite an effect that seems to be rather small. Thus, in the eyes of the majority of the respective stakeholders the merit seems to be big enough to outweigh the disadvantages.

A similar case was already made for another device in soccer before, the goal line technology (see Kolbinger et al., 2015).

In addition, evaluation researches create new knowledge about the underlying phenomena (Stufflebeam & Shinkfield, 2007), which is especially true in this study for the distance set by the referees or the application of the respective set of rules. For the first time, it was shown that there is a discrepancy between the official set of rules for free kicks and its execution on the field, which is also true for the investigation of the referee's estimation of the minimum distance. The respective findings both raise questions concerning the use of the vanishing spray. Users of the device need to be aware that the device controls the compliance of a certain distance, which is in fact not the intended distance for most of the time Oldfather and Fernholz (2009) describe a similar phenomenon concerning the first down marks in American Football). The respective associations also need to be aware, that there is a lack of punishment for minimum distance rule violations, which can not be solved solely by the use of vanishing spray.

7.5 Conclusion

The findings of this study point out the importance of evaluating innovative devices that support the referees in game sports. Based on the objective targets of the respective association the authors showed that the vanishing spray fulfilled its main goal by decreasing the extent of violations at least in some extent. But in addition to that, these evaluations not just generate feedback concerning the new device but also on the underlying phenomena. In this study, the results also indicate a lack of application for the distance rule. Despite a frequent incidence of violations of the minimum distance, none of the investigated free kicks was retaken neither a yellow card was awarded due to this reason. The authors suggest two ways to overcome this discrepancy. On the one hand, the associations could try to increase the awareness of the appropriate punishment of violations of the distance rule. On the other hand, the rule itself could be adapted by changing the extent for the punishment.

Acknowledgments   This chapter is taken from Kolbinger, O. & Link, D. (2016). Vanishing spray reduces extend of rule violations in soccer. *SpringerPlus, 5:1572*. doi:10.1186/s40064-016-3274-2.

8 Prospects

The studies in this book have shown how analytical methods of mathematics and computer science can be used to address performance diagnostic issues in soccer. Looking to the future, it is expected that advances in artificial intelligence, big data, and data mining will both improve existing approaches and lead to completely new applications in sports analytics. The enormous amounts of data in the professional sector allow us to expand our knowledge of performance structure of soccer and put it on an empirically founded basis. This may lead to a new evolutionary stage in training and competition theory for soccer.

The use of data analysis methods will increasingly change the training and competition practice in soccer. This allows us to advance the thesis that competitive advantages are increasingly gained through better and faster information, e. g. training load, opponents' performance status or tactical disposition. This is supported by the fact that the original core area of training science – i.e. essentially the methodology for managing fitness and technical performance prerequisites – has already been optimized over many decades and innovations are comparatively more difficult to produce here than with the provision of information. Even though data analytics has so far been of primary interest to professionals, in the medium term, the methods will also be used in semi-professional and amateur sports.

The high double-digit growth rates at the business end of sport analytics division (Markets and Markets, 2016) will have a direct impact on the employment market and create a large number of new jobs in technology companies, media service providers, and professional clubs. Conferences such as the MIT Sloan Sport Analytics Conference in Boston have meanwhile developed into employment forums for the global market. This creates opportunities for graduates of sports science who have been specifically involved in the field of sports analytics. Against this backdrop, it is surprising that there are hardly any qualification opportunities in sports analytics in Europe so far. Individual topics are certainly dealt with by institutes of training science, sports psychology, performance physiology, sports economics, or business information technology, but there is no integrated program currently available. If we assume that the economic importance of the topic in the USA today will be transferred to the European area in a few years' time, this may be seen as a deficit in education.

An important task for sports science in the coming years will be to follow new developments in information technology and data sciences closely and to examine their potential for improving training and competition practice, as well as for basic research. At the same time, it must also keep an eye on the employment market situation for its graduates. This applies above all with regard to prospects outside a tight academic employment market.

References

Abramitzky, R., Einav, L., Kolkowitz, S. & Mill, R. (2012). On the optimality of line call challenges in professional tennis. *International Economic Review, 53*(3), 939–964. doi:10.1111/j.1468-2354.2012.00706.x

Acar, M., Yapicioglu B., Arikan, N., Yalcin, S., Ates, N. & Ergun, M. (2009). Analysis of goals scored in 2006 World Cup. In T. Reilly & F. Korkusuz (Eds.), *Science & Football VI* (pp. 235–242). Abingdon, Oxon, UK: Routledge.

Alamar, B. C. (2013). *Sports analytics: A guide for coaches, managers, and other decision makers.* New York, NW, USA: Columbia University Press.

Alcock, A. (2010). Analysis of direct free kicks in the women's football World Cup 2007. *European Journal of Sport Science, 10*(4), 279–284. doi:10.1080/17461390903515188

Asmussen, E. & Bøje, O. (1945). Body temperature and capacity for work. *Acta Physiologica Scandinavica, 10*(1), 1–22. doi:10.1111/j.1748-1716.1945.tb00287.x

AT Kearney. (2011). The sports market: Major trends and challenges in an industry full of passion. Retrieved from http://www.atkearney.de/documents/10192/6f46b880-f8d1-4909-9960-cc605bb1ff34

Aughey, R. J., Goodman, C. A. & McKenna, M. J. (2014). Greater chance of high core temperatures with modified pacing strategy during team sport in the heat. *Journal of Science and Medicine in Sport, 17*(1), 113–118. doi:10.1016/j.jsams.2013.02.013

Baeza-Yates, R. & Ribeiro-Neto, B. (1999). *Modern information retrieval.* ACM Press. New York, NY, USA: Addison Wesley.

Bate, R. (1988). Football chance: Tactics and strategy. In Tom Reilly, Adrian Lees, Keith Davids, W. J. Murphy (Ed.), *Science and Football* (pp. 293–301). London, UK: E. & F. N. Spon.

Baumer, B. & Zimbalist, A. (2013). *The sabermetric revolution: Assessing the growth of analytics in baseball.* Philadelphia, PA, USA: University of Pennsylvania Press.

© Springer Fachmedien Wiesbaden GmbH, part of Springer Nature 2018

Bedford, A. & Schembri, A. J. (2006). Back issues: A probability based approach for the allocation of player draft selections in australian rules football. *Journal of Sports Science and Medicine, 5*(4), 509–516.

Beetz, M., von Hoyningen-Huene, N., Kirchlechner, B., Gedikli, S., Siles, F., Durus, M. & Lames, M. (2009). Aspogamo: Automated sports game analysis models. *International Journal of Computer Science in Sport, 8*(1), 1–21.

Bell-Walker, J., McRobert, A., Ford, P. & Williams, M. A. (2006). Quantitative analysis of successful teams at the 2006 World Cup Finals. *Insight – The F.A. Coaches Association Journal, 6*(4), 36–43. doi:10 .1111/j.1748-1716.1987.tb08277.x

Bendiksen, M., Bischoff, R., Randers, M. B., Mohr, M., Rollo, I., Suetta, C., ... Krustrup, P. (2012). The Copenhagen Soccer Test: Physiological response and fatigue development. *Medicine & Science in Sports & Exercise, 44*(8), 1595–1603. doi:10.1249/MSS.0b013e31824cc23b

Bialkowski, A., Lucey, P., Carr, P., Yue, Y., Sridharan, S. & Matthews, I. (2014). Large-scale analysis of soccer matches using spatiotemporal tracking data. In *Proceedings of the 14th IEEE International Conference on Data Mining 2014, Shenzhen, China* (pp. 725–730). Danvers, MA, USA: IEEE. doi:10.1109/ICDM.2014.133

Bialkowski, A., Lucey, P., Carr, P., Denman, S., Matthews, I. & Sridharan, S. (2013). Recognising team activities from noisy data. In *Conference on Computer Vision and Pattern Recognition Workshops, 2013, Portland*. Piscataway, NY, USA: IEEE. doi:10.1109/CVPRW.2013.143

Blatter, S. (2013). Fifa executive committee meeting. opening speech. Zürich, Switzerland.

Blobel, T., Pfab, F., Wanner, P., Haser, C. & Lames, M. (2017). Healthy Reference Patterns (HRP) supporting prevention and rehabilitation process in professional football. In *Proceedings of World Conference on Science in Soccer, 31st May - 2nd June 2017* (p. 183). Reneese: Universitè Reneese 2.

Blomstrand, E. & Essén-Gustavsson, B. (1987). Influence of reduced muscle temperature on metabolism in type I and type II human muscle fibres during intensive exercise. *Acta Physiologica Scandinavica, 131*(4), 569–574. doi:10.1111/j.1748-1716.1987.tb08277.x

Bloomfield, J. R., Polman, R. C. J. & O'Donoghue, P. G. (2005). Effects of score-line on team strategies in FA Premier League Soccer. *Journal of Sports Sciences, 23*(2), 192–193.

Bounfour, A. (2016). *Digital Futures, Digital Transformation, Progress in IS.* Cham: Springer International Publishing. doi:10.1007/978-3-319-23279-9

Bourassa, D. C., McManus, I. C. & Bryden, M. P. (1996). Handedness and eye-dominance: A meta-analysis of their relationship. *Laterality, 1*(1), 5–34.

Bradley, P. S., Carling, C., Archer, D., Roberts, J., Dodds, A., Di Mascio, M., ... Krustrup, P. (2011). The effect of playing formation on high-intensity running and technical profiles in british FA Premier League soccer matches. *Journal of Sports Sciences, 29*(8), 821–830. doi:10.1080/02640414.2011.561868

Bradley, P. S., Carling, C., Diaz, A. G., Hood, P., Barnes, C., Ade, J., ... Mohr, M. (2013). Match performance and physical capacity of players in the top three competitive standards of british professional soccer. *Human Movement Science, 32*(4), 808–821. doi:10.1016/j.humov.2013.06.002

Bradley, P. S. & Noakes, T. D. (2013). Match running performance fluctuations in elite soccer: Indicative of fatigue, pacing or situational influences? *Journal of Sports Sciences, 31*(15), 1627–1638. doi:10.1080/02640414.2013.796062

Buraimo, B., Frick, B., Hickfang, M. & Simmons, R. (2015). The economics of long–term contracts in the footballers' labour market. *Scottish Journal of Political Economy, 62*(1), 8–24. doi:10.1111/sjpe.12064

Cambardella, C. A., Moorman, T. B., Parkin, T. B., Karlen, D. L., Novak, J. M., Turco, R. F. & Konopka, A. E. (1994). Field-scale variability of soil properties in central Iowa soils. *Soil Science Society of America Journal, 58*(5), 1501–1511. doi:10.2136/sssaj1994.03615995005800050033x

Carling, C. (2011). Influence of opposition team formation on physical and skill-related performance in a professional soccer team. *European Journal of Sport Science, 11*(3), 155–164. doi:10.1080/17461391.2010.499972

Carling, C., Williams, A. M. & Reilly, T. (2005). *Handbook of soccer match analysis: A systematic approach to improving performance.* Abingdon, Oxon, UK: Routledge Chapman Hall.

Carling, C. & Bloomfield, J. (2010). The effect of an early dismissal on player work-rate in a professional soccer match. *Journal of Science and Medicine in Sport, 13*(1), 126–128. doi:10.1016/j.jsams.2008.09.004

Carling, C. & Dupont, G. (2011). Are declines in physical performance associated with a reduction in skill-related performance during professional soccer match-play? *Journal of Sports Sciences, 29*(1), 63–71. doi:10.1080/02640414.2010.521945

Carling, C., Wright, C., Nelson, L. J. & Bradley, P. S. (2014). Comment on 'performance analysis in football: A critical review and implications for future research'. *Journal of Sports Sciences, 32*(1), 2–7. doi:10.108 0/02640414.2013.807352

Casal, C. A., Maneiro, R., Ardá, T., Losada, J. L. & Rial, A. (2014). Effectiveness of indirect free kicks in elite soccer. *International Journal of Performance Analysis in Sport, 14*(3), 744–760. doi:10.1080/2474866 8.2014.11868755

Casal, C. A., Maneiro, R., Ardá, T., Losada, J. L. & Rial, A. (2015). Analysis of corner kick success in elite football. *International Journal of Performance Analysis in Sport, 15*(2), 430–451. doi:10.1080/2474866 8.2015.11868805

Castellano, J., Blanco-Villaseñor, A. & Alvarez, D. (2011). Contextual variables and time-motion analysis in soccer. *International Journal of Sports Medicine, 32*(6), 415–421. doi:10.1055/s-0031-1271771

Castellano, J., Casamichana, D. & Lago, C. (2012). The use of match statistics that discriminate between successful and unsuccessful soccer teams. *Journal of Human Kinetics, 31*, 139–147. doi:10.2478/v10078-012-00 15-7

Castells, M. (1996). *The network society.* Oxford, UK: Blackwell.

Cervone, D., DAmour, A., Bornn, L. & Goldsberry, K. (2016). A multiresolution stochastic process model for predicting basketball possession outcomes. *Journal of the American Statistical Association, 111*(514), 585–599. doi:10.1080/01621459.2016.1141685

Chen, H., Chiang, R. H. L. & Storey, V. C. (2012). Business intelligence and analytics: From big data to big impact. *MIS Q, 36*(4), 1165–1188.

Cohen, J. (1992). A power primer. *Psychological Bulletin, 112*(1), 155–159. doi:10.1037/0033-2909.112.1.155

Coleman, B. J. (2012). Identifying the 'players' in sports analytics research. *Interfaces, 42*(2), 109–118. doi:10.1287/inte.1110.0606

Collet, C. (2013). The possession game? A comparative analysis of ball retention and team success in European and international football, 2007–2010. *Journal of Sports Sciences, 31*(2), 123–136. doi:10.1080/0 2640414.2012.727455

Collins, H. & Evans, R. (2008). You cannot be serious! Public understanding of technology with special reference to 'Hawk-Eye'. *Public Understanding of Science, 17*(3), 283–308. doi:10.1177/0963662508093370

Collins, H. & Evans, R. (2012). Sport-decision aids and the 'CSI-effect': Why cricket uses hawk-eye well and tennis uses it badly. *Public Understanding of Science, 21*(8), 904–921. doi:10.1177/0963662511407991

Cooper, G. & Herskovits, E. (1992). A bayesian method for the induction of probabilistic networks from data. *Machine learning, 9*(4), 309–347. doi:10.1007/BF00994110

Cordes, O., Lamb, P. F. & Lames, M. (2012). Concepts and methods for strategy building and tactical adherence: A case study in football. *International Journal of Sports Science & Coaching, 7*(2), 241–254. doi:10.1260/1747-9541.7.2.241

Cramér, H. (1946). *Mathematical methods of statistics*. Princeton, NJ, USA: Princeton University Press.

Cronbach, L. (1963). Course improvement through evaluation. *Teachers College Record, 64*, 672–683. doi:10.1007/978-94-009-6669-7_6

D'Agostino, F. (1995). The ethos of the game. In W. J. Moran & K. V. Meier (Eds.), *Philosophic Inquiry in Sport* (pp. 48–49). Champaign, IL, USA: Human Kinetics. doi:10.1080/00948705.1981.9714372

Davenport, T. H. & Harris, J. G. (2007). *Competing on analytics: The new science of winning*. Brighton, MA, USA: Harvard Business Press.

Deutsche Fußball Liga. (2014). *Definitionskatalog Offizielle Spieldaten (Definitions for Official Gama Data)*. Frankfurt: Deutsche Fußball Liga.

Deutscher, C., Dimant, E. & Humphreys, B. R. (2017). Match fixing and sports betting in football: Empirical evidence from the German Bundesliga. Retrieved from https://ssrn.com/abstract=2910662

Di Prampero, P. E., Fusi, S., Sepulcri, L., Morin, J. B., Belli, A. & Antonutto, G. (2005). Sprint running: A new energetic approach. *Journal of Experimental Biology, 208*(14), 2809–2816. doi:10.1242/jeb.01700

Di Salvo, V., Pigozzi, F., González-Haro, C., Laughlin, M. S. & De Witt, J. K. (2013). Match performance comparison in top English soccer leagues. *International Journal of Sports Medicine, 34*(6), 526–532. doi:10.1055/s-0032-1327660

Dohmen, T. & Sauermann, J. (2015). Referee bias. *Journal of Economic Surveys*. doi:10.1111/joes.12106

D'Orazio, T., Leo, M., Spagnolo, P., Nitti, M., Mosca, N. & Distante, A. (2009). A visual system for real time detection of goal events during soccer matches. *Computer Vision and Image Understanding, 113*(5), 622–632. doi:10.1016/j.cviu.2008.01.010

Doubt, T. J. (1991). Physiology of exercise in the cold. *Sports Medicine (Auckland, N.Z.) 11*(6), 367–381. doi:10.2165/00007256-199111060-00003

Drust, B., Rasmussen, P., Mohr, M., Nielsen, B. & Nybo, L. (2005). Elevations in core and muscle temperature impairs repeated sprint performance. *Acta Physiologica Scandinavica, 183*(2), 181–190. doi:10.1111/j.1365-201X.2004.01390.x

Drust, B., Atkinson, G. & Reilly, T. (2007). Future perspectives in the evaluation of the physiological demands of soccer. *Sports Medicine, 37*(9), 783–805. doi:10.2165/00007256-200737090-00003

Duarte, R., Araújo, D., Correia, V. & Davids, K. (2012). Sports teams as superorganisms. *Sports Medicine, 42*(8), 633–642. doi:10.1007/BF03262285

Duch, J., Waitzman, J. S. & Amaral, L. A. N. (2010). Quantifying the performance of individual players in a team activity. *PLOS ONE, 5*(6), e10937. doi:10.1371/journal.pone.0010937

Dvorak, J. & Racinais, S. (2010). Training and playing football in hot environments. *Scandinavian Journal of Medicine & Science in Sports, 20 Suppl 3*, 4–5. doi:10.1111/j.1600-0838.2010.01203.x

Dvorak, J., Junge, A., Chomiak, J., Graf-Baumann, T., Peterson, L., Rosch, D. & Hodgson, R. (2000). Risk factor analysis for injuries in football players. *The American Journal of Sports Medicine, 28*(5_suppl), 69–74. doi:10.1177/28.suppl5s-69

Edwards, A. M. & Noakes, T. D. (2009). Dehydration; cause of fatigue or sign of pacing in elite soccer? *Sports Medicine, 39*(1), 1–13. doi:10.2165/00007256-200939010-00001

Ekblom, B. (1986). Applied physiology of soccer. *Sports Medicine, 3*(1), 50–60. doi:10.2165/00007256-198603010-00005

Ely, M. R., Martin, D. E., Cheuvront, S. N. & Montain, S. J. (2008). Effect of ambient temperature on marathon pacing is dependent on runner ability. *Medicine and Science in Sports and Exercise, 40*(9), 1675–1680. doi:10.1249/MSS.0b013e3181788da9

Ensum, J., Williams, M. & Grant, A. (2000). Analysis of attacking set plays in Euro 2000. *Insight, 4*(1), 36–39.

FIFA. (2014). *FIFA World Cup semi-final match, Brazil vs Germany match statistics.* Retrieved from http://www.fifa.com/worldcup/matches/round=255955/match=300186474/statistics.html

Fleiss, J. L. (1971). Measuring nominal scale agreement among many raters. *Psychological Bulletin, 76*(5), 378–382. doi:10.1037/h0031619

Folgado, H., Duarte, R., Fernandes, O., Sampaio, J. & Haddad, J. M. (2014). Competing with lower level opponents decreases intra-team movement synchronization and time-motion demands during pre-season soccer matches. *PLOS ONE, 9*(5), e97145. doi:10.1371/journal.pone.0097145

Fonseca, S., Milho, J., Travassos, B. & Araújo, D. (2012). Spatial dynamics of team sports exposed by voronoi diagrams. *Human Movement Science, 31*(6), 1652–1659. doi:10.1016/j.humov.2012.04.006

Frencken, W., Poel, H. d., Visscher, C. & Lemmink, K. (2012). Variability of inter-team distances associated with match events in elite-standard soccer. *Journal of Sports Sciences, 30*(12), 1207–1213. doi:10.1080/02640414.2012.703783

Fullerton, H. S. (1912). The inside game: The science of baseball. *The American Magazine, 70*, 2–13.

Gabbett, T. J. & Domrow, N. (2007). Relationships between training load, injury, and fitness in sub-elite collision sport athletes. *Journal of Sports Sciences, 25*(13), 1507–1519. doi:10.1016/j.jsams.2010.12.002

Gama, J., Passos, P., Davids, K., Relvas, H., Ribeiro, J., Vaz, V. & Dias, G. (2014). Network analysis and intra-team activity in attacking phases of professional football. *International Journal of Performance Analysis in Sport, 14*(3), 692–708.

González-Alonso, J., Calbet, J. A. & Nielsen, B. (1998). Muscle blood flow is reduced with dehydration during prolonged exercise in humans. *The Journal of Physiology, 513*(3), 895–905. doi:10.1111/j.1469-7793.1998.895ba.x

Grant, A. G., Williams, A. M., Reilly, T. & Borrie, T. (1999). Analysis of the goals scored in the 1998 World Cup. *Journal of Sports Sciences, 17*(10), 826–827.

Gregson, W., Drust, B., Atkinson, G. & Salvo, V. D. (2010). Match-to-match variability of high-speed activities in premier league soccer. *International Journal of Sports Medicine, 31*(4), 237–242. doi:10.1055/s-0030-1247546

Grunz, A., Memmert, D. & Perl, J. (2012). Tactical pattern recognition in soccer games by means of special self-organizing maps. *Human Movement Science, 31*(2), 334–343. doi:10.1016/j.humov.2011.02.008

Gudmundsson, J. & Wolle, T. (2014). Football analysis using spatio-temporal tools. *Computers, Environment and Urban Systems, 47*, 16–27. doi:10.1016/j.compenvurbsys.2013.09.004

Hakes, J. K. & Sauer, R. D. (2006). An economic evaluation of the moneyball hypothesis. *The Journal of Economic Perspectives, 20*(3), 173–185. doi:10.1257/089533006780387535

Hargreaves, M. & Febbraio, M. (1998). Limits to exercise performance in the heat. *International Journal of Sports Medicine, 19*(2), S115–116. doi:10.1055/s-2007-971973

Harper, L. D., West, D. J., Stevenson, E. & Russell, M. (2014). Technical performance reduces during the extra-time period of professional soccer match-play. *PLOS ONE, 9*(10), e110995. doi:10.1371/journal.pone.01 10995

Harrop, K. & Nevill, A. (2014). Performance indicators that predict success in an british professional League One soccer team. *International Journal of Performance Analysis in Sport, 14*(3), 907–920. doi:10.108 0/24748668.2014.11868767

Helsen, W., Gilis, B. & Weston, M. (2006). Errors in judging 'offside' in association football: Test of the optical error versus the perceptual flash-lag hypothesis. *Journal of Sports Sciences, 24*(5), 521–528. doi:1 0.1080/02640410500298065

Hernández Moreno, J., Gómez Rijo, A., Castro, U., González Molina, A., Quiroga, M. E. & González Romero, F. (2011). Game rhythm and stoppages in soccer. A case study from Spain. *Journal of Human Sport & Exercise, 6*(4), 594–602. doi:10.4100/jhse.2011.64.03

Hoernig, M., Link, D., Herrmann, M., Radig, B. & Lames, M. (2016). Detection of individual ball possession in soccer. In P. Chung, A. Soltoggio, C. W. Dawson, Q. Meng & M. Pain (Eds.), *Proceedings of the 10th International Symposium on Computer Science in Sports (ISCSS)* (pp. 103–107). Advances in Intelligent Systems and Computing. Cham: Springer International Publishing. doi:10.1007/978-3-319-24560-7_13

Hughes, M. & Bartlett, R. (2002). The use of performance indicators in performance analysis. *Journal of Sports Sciences, 20*(10), 739–754. doi:1 0.1080/026404102320675602

Hughes, M., Dawkins, N., David, R. & Mills, J. (1998). The perturbation effect and goal opportunities in soccer. *Journal of Sports Sciences, 16*(1), 20–21.

Hughes, M. & Franks, I. (2005). Analysis of passing sequences, shots and goals in soccer. *Journal of Sports Sciences, 23*(5), 509–514. doi:10.10 80/02640410410001716779

IBM. (2016). Watson iot and sports: Change the game. Retrieved from https://www.ibm.com/blogs/internet-of-things/watson-iot-sports/

IFAB. (2016). *Laws of the game 2016/17.* Zurich, Switzerland: The International Football Association Board (IFAB).

Intel. (2017). The digitization of sports. Retrieved from https://newsroom. intel.com/press-kits/digitization-of-sports/

Isaaks, E. H. & Srivastava, R. M. (1989). *An introduction to applied geostatistics.* New York, NY, USA: Oxford University Press.

James, N., Rees, G. D., Griffin, E., Barter, P., Taylor, J., Heath, L. & Vučković, G. (2012). Analysing soccer using perturbation attempts. *Journal of Human Sport and Exercise, 7*(2), 413–420. doi:10.4100/jhse.2012.72.07

Jinshan, X., Xiaoke, C., Yamanaka, K. & Matsumoto, M. (1993). Analysis of the goals in the 14th World Cup. In T. Reilly, J. Clarys & A. Stibbe (Eds.), *Science and Football II* (pp. 203–205). London, UK: E. & F. Spon.

Johnson, J. G. (2006). Cognitive modeling of decision making in sports. *Psychology of Sport and Exercise, 7*(6), 631–652. doi:10.1016/j.psychsport.2006.03.009

Jones, P. D., James, N. & Mellalieu, S. D. (2004). Possession as a performance indicator in soccer. *International Journal of Performance Analysis in Sport, 4*(1), 98–102. doi:10.1080/24748668.2004.11868295

Kang, C.-h., Hwang, J.-r. & Li, K.-j. (2006). Trajectory analysis for soccer players. In *Sixth IEEE International Conference on Data Mining - Workshops (ICDMW'06, hong kong)* (pp. 377–381). Piscataway, NY, USA: IEEE. doi:10.1109/ICDMW.2006.160

Kempe, M., Vogelbein, M., Memmert, D. & Nopp, S. (2014). Possession vs. direct play: Evaluating tactical behavior in elite soccer. *International Journal of Sports Science, 4*(6A), 35–41. doi:10.5923/s.sports.201401.05

Kerwin, D. G. & Bray, K. (2006). Measuring and modelling the goalkeeper's diving envelope in a penalty kick. In E. Moritz & S. Haake (Eds.), *The Engineering of Sport 6* (pp. 321–326). Hamburg: Springer. doi:10.1007/978-0-387-46050-5_57

Knudsona, D. & Klukab, D. A. (1997). The impact of vision and vision training on sport performance. *Journal of Physical Education, Recreation and Dance, 68*(4), 17–24. doi:10.1080/07303084.1997.10604922

Kolbinger, O. & Link, D. (2016). Vanishing spray reduces extend of rule violations in soccer. *SpringerPlus, 5:1572.* doi:10.1186/s40064-016-3274-2

Kolbinger, O., Linke, D., Link, D. & Lames, M. (2015). Do we need goal line technology in soccer or could video proof be a more suitable choice.

In J. Cabri, J. Barreiros & P. Pezarat-Correia (Eds.), *Sports Science Research and Technology Support: Second International Congress, ic-SPORTS 2014, Revised Selected Papers* (pp. 107–118). Cham: Springer International Publishing. doi:10.1007/978-3-319-25249-0_8

Komar, J. (2015). On the use of tracking data to support coaches in professional footbal. In R. Duarte, B. Eskofier, M. Rumpf & J. Wiemeyer (Eds.), *Modeling and Simulation of Sport Games, Sport Movements, and Adaptations to Training (Dagstuhl Seminar 15382)* (p. 42). Dagstuhl, Germany.

Lago, C. & Martín, R. (2007). Determinants of possession of the ball in soccer. *Journal of Sports Sciences, 25*(9), 969–974. doi:10.1080/02640410600944626

Lago-Ballesteros, J. & Lago-Peñas, C. (2010). Performance in team sports: Identifying the keys to success in soccer. *Journal of Human Kinetics, 25.* doi:10.2478/v10078-010-0035-0

Lago-Peñas, C. & Dellal, A. (2010). Ball possession strategies in elite soccer according to the evolution of the match-score: The influence of situational variables. *Journal of Human Kinetics, 25.* doi:10.2478/v10078-010-0036-z

Lago-Peñas, C. & Lago-Ballesteros, J. (2011). Game location and team quality effects on performance profiles in professional soccer. *Journal of Sports Science and Medicine, 10*(3), 465–471.

Lames, M. & McGarry, T. (2007). On the search for reliable performance indicators in game sports. *International Journal of Performance Analysis in Sport, 7*(1), 62–79.

Lawlor, J., Low, D., Taylor, S. & Williams, A. M. (2003). The FIFA World Cup 2002: An analysis of successful versus unsuccessful teams. *Journal of Sports Sciences, 22*(6), 500–520.

Le, H. M., Carr, P., Yue, Y. & Lucey, P. (2017). Data-driven ghosting using deep imitation learning. In *Proceeding of the 11th MIT Sloan Sports Analytics Conference 2017, Boston.* Boston, MA, USA: MIT.

Lepioufle, J.-M., Leblois, E. & Creutin, J.-D. (2012). Variography of rainfall accumulation in presence of advection. *Journal of Hydrology, 464-465,* 494–504. doi:10.1016/j.jhydrol.2012.07.041

Leser, R., Baca, A. & Ogris, G. (2011). Local positioning systems in (game) sports. *Sensors, 11*(10), 9778–9797. doi:10.3390/s111009778

Lewis, M. (2003). *Moneyball: The art of winning an unfair game.* New York, NW, USA: W. W. Norton & Company.

Link, D. (2014). Using of invasion profiles as a performance indicator in soccer. In I. Heazelwood (Ed.), *Proceeding of International Association of Computer Science in Sports 2014 Conference, IACSS 2014*. Darwin, Australia: Charles Darwin University.

Link, D. & de Lorenzo, M. F. (2016). Seasonal pacing – Match importance affects activity in professional soccer. *PLOS ONE*, *11*(6), e0157127. doi:10.1371/journal.pone.0157127

Link, D. & Hoernig, M. (2017). Individual ball possession in soccer. *PLOS ONE*, *7*(5). doi:10.1371/journal.pone.0179953

Link, D., Kolbinger, O., Weber, H. & Stöckl, M. (2016). A topography of free kicks in soccer. *Journal of Sports Sciences*, *34*(24), 2312–2320. doi:10.1080/02640414.2016.1232487

Link, D. & Lames, M. (2014). An introduction to sport informatics. In *Computer Science in Sport: Research and Practice* (pp. 1–17). New York, NY, USA: Routledge. doi:10.4324/9781315881782

Link, D., Lang, S. & Seidenschwarz, P. (2016). Real time quantification of dangerousity in football using spatiotemporal tracking data. *PLOS ONE*, *11*(12), e0168768. doi:10.1371/journal.pone.0168768

Link, D. & Weber, H. (2017). Effect of ambient temperature on pacing in soccer depends on skill level. *Journal of Strength and Conditioning Research*, *31*(7), 1766–1770. doi:10.1519/JSC.0000000000001013

López-Botella, M. & Palao, J. M. (2007). Relationship between laterality of foot strike and shot zone on penalty efficacy in specialist penalty takers. *International Journal of Performance Analysis in Sport*, *7*(3), 26–36. doi:10.1080/24748668.2007.11868407

Lucey, P., Bialkowski, A., Monfort, M., Carr, P. & Matthews, I. (2015). Quality vs quantity: Improved shot prediction in soccer using strategic features from spatiotemporal data. In *Proceeding of MIT Sloan Sports Analytics Conference, 2015*. Boston, MA, USA: MIT Sloan Group.

Lucey, P., Oliver, D., Carr, P., Roth, J. & Matthews, I. (2013). Assessing team strategy using spatiotemporal data. In R. L. Grossman, R. Uthurusamy, I. Dhillon & Y. Koren (Eds.), *19th ACM SIGKDD International Conference, Chicago, 2013* (p. 1366). New York, USA: ACM. doi:10.1145/2487575.2488191

Mackenzie, R. & Cushion, C. (2013). Performance analysis in football: A critical review and implications for future research. *Journal of Sports Sciences*, *31*(6), 639–676. doi:10.1080/02640414.2012.746720

Markets and Markets. (2016). Sports analytics market - global forecast to 2021. Retrieved from www.marketsandmarkets.com/Market-Reports/sports-analytics-market-35276513.html

Markie, P. (2015). Rationalism vs. empiricism. In Edward D. Zalta (Ed.), *Stanford Encyclopedia of Philosophy*. ePrint. Retrieved from http://plato.stanford.edu/archives/sum2015/entries/rationalism-empiricism/

Mather, G. (2008). Perceptual uncertainty and line-call challenges in professional tennis. *Proceedings of the Royal Socienty B - Biological sciences, 275*(1643), 1645–1651. doi:10.1098/rspb.2008.0211

Microsoft. (2017). Sports performance platform puts data into play – and action – for athletes and teams. Retrieved from https://blogs.microsoft.com/blog/2017/06/27/sports-performance-platform-puts-data-play-action-athletes-teams

Miller, T. W. (2015). *Sports analytics and data science: Winning the game with methods and models (FT Press Analytics)*. Old Tappan, NJ, USA: Pearson Education.

Mohr, M. & Krustrup, P. (2013). Heat stress impairs repeated jump ability after competitive elite soccer games. *Journal of Strength and Conditioning Research, 27*(3), 683–689. doi:10.1097/JSC.0b013e31825c3266

Mohr, M., Krustrup, P. & Bangsbo, J. (2005). Fatigue in soccer: A brief review. *Journal of Sports Sciences, 23*(6), 593–599. doi:10.1080/02640410400021286

Mohr, M., Nybo, L., Grantham, J. & Racinais, S. (2012). Physiological responses and physical performance during football in the heat. *PLOS ONE, 7*(6), e39202. doi:10.1371/journal.pone.0039202

Montain, S. J., Ely, M. R. & Cheuvront, S. N. (2007). Marathon performance in thermally stressing conditions. *Sports Medicine, 37*(4-5), 320–323. doi:10.2165/00007256-200737040-00012

Morris, C. (1977). The most important points in tennis. *Optimal strategies in sports, 5*, 131–140.

Morya, E., Bigatao, H., Lees, A. & Ranvaud, R. (2005). Evolving penalty kick strategies: World Cup and club matches, 2000-2002. In T. Reilly, J. Cabri & D. Araújo (Eds.), *Science and Football V* (pp. 237–242). London, UK: E & FN Spon London.

Mugglestone, C., Morris, J. G., Saunders, B. & Sunderland, C. (2013). Half-time and high-speed running in the second half of soccer. *International Journal of Sports Medicine, 34*(6), 514–519. doi:10.1055/s-0032-1327647

Mujika, I. & Padilla, S. (2003). Scientific bases for precompetition tapering strategies. *Medicine and Science in Sports and Exercise, 35*(7), 1182–1187. doi:10.1249/01.MSS.0000074448.73931.11

Njororai, W. W. (2013). Analysis of goals scored in the 2010 world cup soccer tournament held in South Africa. *Journal of Physical Education and Sport, 13*(1), 6–13.

Nybo, L., Møller, K., Volianitis, S., Nielsen, B. & Secher, N. H. (2002). Effects of hyperthermia on cerebral blood flow and metabolism during prolonged exercise in humans. *Journal of Applied Physiology, 93*(1), 58–64. doi:10.1152/japplphysiol.00049.2002

Nybo, L. & Secher, N. H. (2004). Cerebral perturbations provoked by prolonged exercise. *Progress in Neurobiology, 72*(4), 223–261. doi:10.101 6/j.pneurobio.2004.03.005

Oberstone, J. (2009). Differentiating the top english premier league football clubs from the rest of the pack: Identifying the keys to success. *Journal of Quantitative Analysis in Sports, 5*(3). doi:10.2202/1559-0410.1183

Oldfather, C. M. & Fernholz, M. M. (2009). Comparative procedure on a sunday afternoon: Instant replay in the NFL as a process of appellate review. *Indiana Law Review, 43*(1), 45–78. doi:10.18060/4027

Olsen, E., Larsen, O., Reilly, T., Bangsbo, J. & Hughes, M. (1995). Use of match analysis by coaches. In J. Bangsbo, T. Reilly & A. M. Williams (Eds.), *Science and Football III* (pp. 209–220). Abington, Oxon, UK: Taylor & Francis.

Osgnach, C., Poser, S., Bernardini, R., Rinaldo, R. & Di Prampero, P. E. (2010). Energy cost and metabolic power in elite soccer: A new match analysis approach. *Medicine & Science in Sports & Exercise, 42*(1), 170–178. doi:10.1249/MSS.0b013e3181ae5cfd

Özgünen, K. T., Kurdak, S. S., Maughan, R. J., Zeren, C., Korkmaz, S., Yazici, Z., ... Dvorak, J. (2010). Effect of hot environmental conditions on physical activity patterns and temperature response of football players. *Scandinavian Journal of Medicine & Science in Sports, 20*(3), 140–147. doi:10.1111/j.1600-0838.2010.01219.x

Panchanathan, S., Chakraborty, S., McDaniel, T., Bunch, M., O'Connor, N., Little, S., ... Marsden, M. (2016). Smart stadium for smarter living: Enriching the fan experience. In *International Symposium on Multimedia (ISM) 2016* (pp. 152–157). San Jose, CA, USA: IEEE. doi:10 .1109/ISM.2016.0037

Parsons, C. A., Sulaeman, J., Yates, M. C. & Hamermesh, D. S. (2011). Strike three: Discrimination, incentives, and evaluation. *American Economic Review, 101*(4), 1410–35. doi:10.1257/aer.101.4.1410

Paul, D. J., Bradley, P. S. & Nassis, G. P. (2015). Factors affecting match running performance of elite soccer players: Shedding some light on the complexity. *International Journal of Sports Physiology and Performance, 10*(4), 561–519. doi:10.1123/IJSPP.2015-0029

Pebesma, E. J. (2004). Multivariable geostatistics in S: The gstat package. *Computers & Geosciences, 30*(7), 683–691. doi:10.1016/j.cageo.2004.03.012

Pendergast, D. R. (1988). The effect of body cooling on oxygen transport during exercise. *Medicine and Science in Sports and Exercise, 20*(5 Suppl), S171–176.

Pollard, R., Ensum, J. & Taylor, S. (2004). Estimating the probability of a shot resulting in a goal: The effects of distance, angle and space. *International Journal of Soccer and Science, 2*(1), 50–55.

Pope, B. R. & Pope, N. G. (2015). Own-nationality bias: Evidence from UEFA Champions League football referees. *Economic Inquiry, 53*(2), 1292–1304. doi:10.1111/ecin.12180

Pratas, J., Volossovitch, A. & Ferreira, A. P. (2012). The effect of situational variables on teams' performance in offensive sequences ending in a shot on goal. a case study. *The Open Sports Sciences Journal, 5*(5), 193–199. doi:10.2174/1875399X01205010193

Psiuk, R., Seidl, T., Strauß, W. & Bernhard, J. (2014). Analysis of goal line technology from the perspective of an electromagnetic field based approach. *Procedia Engineering, 72*, 279–284. doi:10.1016/j.proeng.2014.06.050

Qing, W., Hengshu, Z., Wei, H., Zhiyong, S. & Yuan, Y. (2015). Discerning tactical patterns for professional soccer teams: An enhanced topic model with applications. In *Proceedings of the 21st ACM SIGKDD International Conference on Knowledge Discovery and Data Mining, Sydney, NSW, Australia* (pp. 2197–2206). New York, NY, USA: ACM. doi:10.1145/2783258.2788577

Rampinini, E., Coutts, A. J., Castagna, C., Sassi, R. & Impellizzeri, F. M. (2007). Variation in top level soccer match performance. *International Journal of Sports Medicine, 28*(12), 1018–1024. doi:10.1055/s-2007-965158

Rein, R. & Memmert, D. (2016). Big data and tactical analysis in elite soccer: Future challenges and opportunities for sports science. *SpringerPlus, 5*(1), 1410. doi:10.1186/s40064-016-3108-2

Reis, A. P., da Silva, E. F., Sousa, A. J., Patinha, C. & Fonseca, E. C. (2007). Spatial patterns of dispersion and pollution sources for arsenic at Lousal mine, Portugal. *International Journal of Environmental Health Research, 17*(5), 335–349. doi:10.1080/09603120701628412

Research and Markets. (2016). Worldwide Sports Analytics Market (2016-2022). Retrieved from www.researchandmarkets.com/research/gpr3f w/worldwide_sports

Ribeiro, J., Silva, P., Duarte, R., Davids, K. & Garganta, J. (2017). Team sports performance analysed through the lens of social network theory: Implications for research and practice. *Sports Medicine*, 1–8. doi:10.1 007/s40279-017-0695-1

Rickey, B. (1954). Goodbye to some old baseball ideas. *Life, August 2*, 78–89.

Rock, R., Als, A., Gibbs, P. & Hunte, C. (2013). The 5th umpire: Automating crickets edge detection system. *Journal of Systemics, Informatics & Cybernetics, 11*(1), 4–9.

Samal, A. R., Sengupta, R. R. & Fifarek, R. H. (2011). Modelling spatial anisotropy of gold concentration data using GIS-based interpolated maps and variogram analysis: Implications for structural control of mineralization. *Journal of Earth System Science, 120*(4), 583–593. doi:10.1007/s12040-011-0091-4

Sands, W. A., Kavanaugh, A. A., Murray, S. R., McNeal, J. R. & Jemni, M. (2016). Modern techniques and technologies applied to training and performance monitoring. *International Journal of Sports Physiology and Performance*, 1–29. doi:10.1123/ijspp.2016-0405

SAP. (2015). Unveils sap sports one solution for soccer. Retrieved from http://news.sap.com/sap-unveils-sap-sports-one-solution-for-soccer/

Särkkä, S. (2013). *Bayesian filtering and smoothing*. Cambridge, UK: Cambridge University Press.

Schilling, M. F. (1994). The importance of a game. *Mathematics Magazine, 67*(4), 282–288. doi:10.2307/2690849

Scriven, M. (1991). *Evaluation thesaurus*. Newbury Park, CA, USA: Sage.

Shah, F. A., Kretzer, M. & Mädche, A. (2015). Designing an analytics platform for professional sports teams. In *36th International Conference on Information Systems, Fort Worth 2015* (pp. 1–20). Atlanta, GA, USA: Association for Information Systems.

Siegle, M. & Lames, M. (2012). Game interruptions in elite soccer. *Journal of Sports Sciences, 30*(7), 619–624. doi:10.1080/02640414.2012.667877

Siegle, M., Stevens, T. & Lames, M. (2013). Design of an accuracy study for position detection in football. *Journal of Sports Sciences, 31*(2), 166–172. doi:10.1080/02640414.2012.723131

Silva, J. R., Magalhães, J. F., Ascensão, A. A., Oliveira, E. M., Seabra, A. F. & Rebelo, A. N. (2011). Individual match playing time during the season affects fitness-related parameters of male professional soccer players. *The Journal of Strength & Conditioning Research, 25*(10), 2729–2739. doi:10.1519/JSC.0b013e31820da078

Silva, J. R., Magalhães, J., Ascensão, A., Seabra, A. F. & Rebelo, A. N. (2013). Training status and match activity of professional soccer players throughout a season. *The Journal of Strength & Conditioning Research, 27*(1), 20–30. doi:10.1519/JSC.0b013e31824e1946

Simmons, R. (2007). Overpaid athletes? Comparing american and european football. *WorkingUSA, 10*(4), 457–471. doi:10.1111/j.1743-4580.2007 .00176.x

Sink, K. R., Thomas, T. R., Araujo, J. & Hill, S. F. (1989). Fat energy use and plasma lipid changes associated with exercise intensity and temperature. *European Journal of Applied Physiology and Occupational Physiology, 58*(5), 508–513. doi:10.1007/BF02330705

Spearman, W., Basye, A., Dick, G., Hotovy, R. & Pop, P. (2017). Physics-based modeling of pass probabilities in soccer. In *Proceeding of the 11th MIT Sloan Sports Analytics Conference 2017, Boston*. Boston, MA, USA: MIT.

Stöckl, M., Lamb, P. F. & Lames, M. (2011). The ISOPAR method: A new approach to performance analysis in golf. *Journal of Quantitative Analysis in Sports, 7*(1). doi:10.2202/1559-0410.1289

Stöckl, M., Lamb, P. F. & Lames, M. (2012). A model for visualizing difficulty in golf and subsequent performance rankings on the PGA tour. *International Journal of Golf Science, 1*(1), 10–24. doi:10.1123/ijgs.1 .1.10

Stöckl, M. & Lames, M. (2012). Creating a continuous topography of performance from discrete sports actions. *Mathematical Modelling, 7*(1), 814–819. doi:10.3182/20120215-3-AT-3016.00144

Stöckl, M. & Morgan, S. (2013). Visualization and analysis of spatial characteristics of attacks in field hockey. *International Journal of Performance Analysis in Sport, 13*(1), 160–178. doi:10.1080/24748668.2013.11 868639

Stufflebeam, D. L. (1983). The CIPP model for program evaluation. In *Evaluation Models 6* (pp. 117–141). Evaluation in Education and Human Services. doi: 10.1007/978-94-009-6669-7_7. Dordrecht, Netherlands: Springer Netherlands. doi:10.1007/978-94-009-6669-7_7

Stufflebeam, D. L. & Shinkfield, A. (2007). *Evaluation theory, models, and applications*. San Francisco, CA, USA: Jossey-Bass.

Taylor, J. B., James, N. & Mellalieu, S. D. (2005). Notational analysis of corner kicks in british premier league soccer. In T. Reilly, J. Cabri &

D. Araújo (Eds.), *Science and Football V* (pp. 229–234). London, UK: E & FN Spon London.

Tenga, A., Holme, I., Ronglan, L. T. & Bahr, R. (2010). Effect of playing tactics on achieving score-box possessions in a random series of team possessions from Norwegian professional soccer matches. *Journal of Sports Sciences, 28*(3), 245–255. doi:10.1080/02640410903502766

Varley, M. C., Elias, G. P. & Aughey, R. J. (2012). Current match-analysis techniques' underestimation of intense periods of high-velocity running. *International Journal of Sports Physiology and Performance, 7*(2), 183–185.

Vecer, J. (2014). Crossing in soccer has a strong negative impact on scoring: Evidence from the british Premier League the German Bundesliga and the World Cup 2014. *Available at SSRN 2225728*. doi:10.2139/ssrn.2 225728

Vilar, L., Araújo, D., Davids, K. & Button, C. (2012). The role of ecological dynamics in analysing performance in team sports. *Sports Medicine, 42*(1), 1–10. doi:10.2165/11596520-000000000-00000

Wei, X., Sha, L., Lucey, P., Morgan, S. & Sridharan, S. (2013). Large-scale analysis of formations in soccer. In *2013 International Conference on Digital Image Computing: Techniques and Applications (DICTA), Hobart, Tasmania* (pp. 1–8). Danvers, MA, USA: IEEE. doi:10.1109 /DICTA.2013.6691503

Wright, C., Atkins, S., Jones, B. & Todd, J. (2013). The role of performance analysts within the coaching process: Performance Analysts Survey 'The role of performance analysts in elite football club settings'. *International Journal of Performance Analysis in Sport, 13*(1), 240–261. doi:10.1080/24748668.2013.11868645

Wright, M. B. (2009). 50 years of OR in sport. *Journal of the Operational Research Society, 60*(1), 161–168. doi:10.1057/jors.2008.170

Yiannakos, A. & Armatas, V. (2006). Evaluation of the goal scoring patterns in European Championship in Portugal 2004. *International Journal of Performance Analysis in Sport, 6*(1), 178–188. doi:10.1080/24748668 .2006.11868366

Yiannis, M. (2014). Analysis of goals scored in the 2014 World Cup soccer tournament held in Brazil. *International Journal of Sport Studies, 4*(9), 1017–1026.

Printed in the United States
By Bookmasters